AF212914

Birkhäuser

Operator Theory: Advances and Applications
Volume 214

Founded in 1979 by Israel Gohberg

Editors:
Joseph A. Ball (Blacksburg, VA, USA)
Harry Dym (Rehovot, Israel)
Marinus A. Kaashoek (Amsterdam, The Netherlands)
Heinz Langer (Vienna, Austria)
Christiane Tretter (Bern, Switzerland)

Associate Editors:
Vadim Adamyan (Odessa, Ukraine)
Albrecht Böttcher (Chemnitz, Germany)
B. Malcolm Brown (Cardiff, UK)
Raul Curto (Iowa, IA, USA)
Fritz Gesztesy (Columbia, MO, USA)
Pavel Kurasov (Lund, Sweden)
Leonid E. Lerer (Haifa, Israel)
Vern Paulsen (Houston, TX, USA)
Mihai Putinar (Santa Barbara, CA, USA)
Leiba Rodman (Williamsburg, VA, USA)
Ilya M. Spitkovsky (Williamsburg, VA, USA)

Honorary and Advisory Editorial Board:
Lewis A. Coburn (Buffalo, NY, USA)
Ciprian Foias (College Station, TX, USA)
J.William Helton (San Diego, CA, USA)
Thomas Kailath (Stanford, CA, USA)
Peter Lancaster (Calgary, Canada)
Peter D. Lax (New York, NY, USA)
Donald Sarason (Berkeley, CA, USA)
Bernd Silbermann (Chemnitz, Germany)
Harold Widom (Santa Cruz, CA, USA)

Subseries
Linear Operators and Linear Systems
Subseries editors:
Daniel Alpay (Beer Sheva, Israel)
Birgit Jacob (Wuppertal, Germany)
André C.M. Ran (Amsterdam, The Netherlands)

Subseries
Advances in Partial Differential Equations
Subseries editors:
Bert-Wolfgang Schulze (Potsdam, Germany)
Michael Demuth (Clausthal, Germany)
Jerome A. Goldstein (Memphis, TN, USA)
Nobuyuki Tose (Yokohama, Japan)
Ingo Witt (Göttingen, Germany)

Spectral Theory and Analysis

Conference on Operator Theory, Analysis and Mathematical Physics (OTAMP) 2008, Bedlewo, Poland

Jan Janas
Pavel Kurasov
Ari Laptev
Sergei Naboko
Günter Stolz
Editors

Birkhäuser

Editors

Jan Janas
Institut of Mathematics
Polish Academy of Sciences
ul. Sw. Tomasza 30/7
31-027 Krakow, Poland
najanas@cyf-kr.edu.pl

Pavel Kurasov
Department of Mathematics
Lund Institute of Technology
221 00 Lund, Sweden
kurasov@maths.lth.se

Ari Laptev
Department of Mathematics
South Kensington Campus
Imperial College London
London SW7 2AZ, UK
a.laptev@imperial.ac.uk
and
Department of Mathematics
Royal Institute of Technology
100 44 Stockholm, Sweden
laptev@math.kth.se

Sergei Naboko
Department of Mathematical Physics
St. Petersburg State University
Ulyanovskaya 1
198904 St. Petersburg, Russia
naboko@snoopy.phys.spbu.ru

Günter Stolz
Department of Mathematics
University of Alabama
452 Campbell Hall
Birmingham, AL 35294-1170, USA
stolz@math.uab.edu

2010 Mathematical Subject Classification: Primary: 81-06 (81Q10, 81Q12, 81Q15, 81Q20, 81Q80); Secondary: 35-06 (35P05, 35P15, 35P20, 35P30), 47-06 (47B25, 47F05)

ISBN 978-3-0348-0326-7 ISBN 978-3-7643-9994-8 (eBook)
DOI 10.1007/978-3-7643-9994-8

© Springer Basel AG 2011
Softcover re-print of the Hardcover 1st edition 2011
This work is subject to copyright. All rights are reserved, whether the whole or part of the material is concerned, specifically the rights of translation, reprinting, re-use of illustrations, recitation, broadcasting, reproduction on microfilms or in other ways, and storage in data banks. For any kind of use permission of the copyright owner must be obtained.

Cover design: deblik, Berlin

Printed on acid-free paper

Springer Basel AG is part of Springer Science+Business Media

www.birkhauser-science.com

This volume is dedicated to the
memory and achievements of

Mikhail Shlemovich Birman

an outstanding mathematician of the
20th century in Mathematical Physics

Mikhail Shlemovich Birman 1928–2009

Contents

Contents

Introduction

This volume appears as an outcome of the conference Operator Theory, Analysis and Mathematical Physics – OTAMP2008, held at Mathematical Research and Conference Center in Bedlewo near Poznan. The volume contains few review articles as well as original research papers presented at the conference or appeared as a result of inspiring discussions during the meeting.

The titles of conference talks followed the subjects of four special sessions:

- random and quasi-periodic differential operators
- orthogonal polynomials
- Jacobi and CMV matrices
- quantum graphs

All contributions to this volume are devoted to different chapters of operator theory with a focus towards applications in mathematical physics. Several articles are in the area of spectral theory of Schrödinger operators having emphasis on problems with magnetic fields. Another subject well-represented concerns spectral theory for non-self-adjoint problems. Spectral analysis is not restricted to just linear and self-adjoint problems.

This volume is devoted to the memory of Mikhail Shlemovich Birman – one of the most outstanding scientists of the last century. The influence of his ideas on the development of mathematical physics in the whole world and especially in Saint Petersburg will continue for decades, several of the authors contributed to this volume have been his students and will carry over his special attitude towards science to new generations to come.

Preparing this volume we remembered another remarkable mathematical physicist Israel Gohberg who always supported OTAMP conferences by including proceedings into the series *Operator Theory: Advances and Applications* and helping us with selection of outstanding contributors and interesting subjects in operator theory.

We would like to thank the European Science Foundation (ESF) for a generous financial support which allowed to transfer the OTAMP conference into a major event in the area of mathematical physics in 2008. We are grateful to all people working at Mathematical Research and Conference Center in Bedlewo for creating a stimulating scientific atmosphere and help before, during and after the conference.

Birmingham-Krakow-London | May 2010
Lund-St. Petersburg-Stockholm | The Editors

Operator Theory:
Advances and Applications, Vol. 214, 1–20
© 2011 Springer Basel AG

Floquet-Bloch Theory for Elliptic Problems with Discontinuous Coefficients

B.M. Brown, V. Hoang, M. Plum and I.G. Wood

Abstract. We study spectral properties of elliptic problems of order $2m$ with periodic coefficients in L^∞. Our goal is to obtain a Floquet-Bloch type representation of the spectrum in terms of the spectra of associated operators acting on the period cell. Our approach using bilinear forms and operators in H^{-m}-type spaces easily handles discontinuous coefficients and has the merit of being rather direct. In addition, the cell of periodicity is allowed to be unbounded, i.e., periodicity is not required in all spatial directions.

Mathematics Subject Classification (2000). 35J10, 35j30, 35J99, 35P10.

Keywords. Floquet-Bloch, $2m$th-order elliptic, spectral theory.

1. Introduction

One of the most important partial differential operators in quantum physics is the Schrödinger operator

$$-\Delta + V(x), \quad x \in \mathbb{R}^d.$$

In many application areas the potential $V(\cdot)$ is periodic with respect to a lattice in $\mathbb{R}^d$. An extension of this equation to include a magnetic term gives rise to the magnetic Schrödinger operator

$$(-i\nabla - A(x))^2 + V(x), \quad x \in \mathbb{R}^d$$

where now both the potential V and the magnetic potential A are periodic with respect to the underlying lattice. Further examples of elliptic partial differential operators which have periodic coefficients may also be found in the periodic Dirac operator, the fields of periodic acoustics and photonic crystals. In all these cases of periodic coefficients, the main way of studying the spectrum of a suitable self-adjoint realisation is via the so-called Floquet-Bloch theory where essentially the spectral properties of the operator in $\mathbb{R}^d$ are read off from the behaviour on a fundamental cell of periodicity resulting in the well-known band-gap structure. For further information, see [2–5, 7, 8, 10, 11, 13–15] and the references quoted therein.

Often, in particular when the potentials are non-smooth, it is advantageous to study these problems using an associated bilinear form, and this is the approach that we take. We work with an mth-order elliptic Hermitian sesquilinear form on $H^m(\mathbb{R}^d)$ with periodic and bounded coefficients which we also allow to be discontinuous. This is equivalent to studying a selfadjoint operator L (associated with the bilinear form) in the dual space $H^{-m}(\mathbb{R}^d)$. Our motivation for this is found in the study of crystals where in practice often two (or more) materials are combined to form a periodic structure, which results in piecewise constant periodic coefficients. For future investigation of wave-guide properties we will require periodicity of the coefficients only in some spatial directions, i.e., we allow unbounded periodic cells. Hence in general no Bloch waves are available, and we have to use other techniques replacing the usual Bloch wave expansion, e.g., we prove that the Floquet transform (which is known to be an isometric isomorphism between L^2-spaces) is an isometric isomorphism also between H^{-m} spaces (see [10] for further mapping properties of the Floquet transformation).

Our result gives the well-known decomposition of the spectrum of periodic differential operators, developed, e.g., in [10], [14], [4], now also in the case of discontinuous coefficients (including the principal ones) and unbounded periodicity cell. A corresponding result is stated in [6], lacking however a detailed and self-contained proof, which we will give in this paper in a rather direct way.

We shall further show that the spectrum of L coincides with the spectrum of a suitable operator $\widetilde{\mathrm{L}}$ in $L^2(\mathbb{R}^d)$ associated with the bilinear form, which is constructed in a standard way. A direct study of the spectrum of $\widetilde{\mathrm{L}}$ by the "usual" Floquet-Bloch theory in $L^2(\mathbb{R}^d)$ seems to be problematic due to lack of smoothness in the coefficients.

2. Definitions and preliminary results

In the following, $H^m(\mathbb{R}^d)$ will always denote the Sobolev space of functions which are square Lebesgue-integrable over $\mathbb{R}^d$ with square integrable derivatives up to order m. We denote the usual norm by

$$\|u\|^2_{H^m(\mathbb{R}^d)} = \sum_{|\alpha|\leq m} \|D^\alpha u\|^2_{L^2(\mathbb{R}^d)} \, .$$

Let

$$B : H^m(\mathbb{R}^d) \times H^m(\mathbb{R}^d) \to \mathbb{C}$$

be a closed Hermitian sesquilinear form. We write $d = d_1 + d_2$ and use variables $x \in \mathbb{R}^{d_1}$ and $y \in \mathbb{R}^{d_2}$. We assume B is given in the form

$$B[u,v] := \sum_{|\rho|,|\sigma|\leq m} \int_{\mathbb{R}^d} a^{\rho\sigma}(x,y)(D^\rho u)(x,y)\overline{(D^\sigma v)(x,y)}dxdy \tag{2.1}$$

where $a^{\rho\sigma} \in L^\infty(\mathbb{R}^d)$, $a^{\rho\sigma} = \overline{a^{\sigma\rho}}$, and $a^{\rho\sigma}(x,y) = a^{\rho\sigma}(x, y+e_j)$ for any $x \in \mathbb{R}^{d_1}$, $y \in \mathbb{R}^{d_2}$, $j = 1, \ldots, d_2$, where $e_1, \ldots, e_{d_2}$ are the unit vectors[1] in $\mathbb{R}^{d_2}$. Moreover, we assume that the leading coefficients satisfy the ellipticity condition[2]

$$\sum_{|\rho|,|\sigma|=m} a^{\rho\sigma}(x,y)\zeta_\rho\overline{\zeta_\sigma} \geq c \sum_{|\alpha|=m} |\zeta_\alpha|^2 \tag{2.2}$$

for some $c > 0$, all $(x,y) \in \mathbb{R}^d$ and all $\zeta = (\zeta_\alpha)_{|\alpha|=m} \in \mathbb{C}^N$, where $N = \#\{\alpha : |\alpha| = m\}$.

Remark 2.1. This condition, which (for the case of real-valued coefficients) appears, e.g., in [12], is, in general, stronger than the usual strong ellipticity condition

$$\operatorname{Re} \sum_{|\rho|,|\sigma|=m} a^{\rho\sigma}(x,y)\xi^\rho\xi^\sigma \geq c|\xi|^{2m}$$

for all $\xi \in \mathbb{R}^d$ and $(x,y) \in \mathbb{R}^d$. We need this stronger condition, since we want to avoid the assumption of continuity of the leading coefficients $a^{\rho\sigma}$.

Throughout this paper, let $\Omega := \mathbb{R}^{d_1} \times [0,1]^{d_2}$ denote the periodic cell for our problem. We also introduce a bilinear form acting on Ω. Let

$$B_\Omega[u,v] := \sum_{|\rho|,\,|\sigma|\leq m} \int_\Omega a^{\rho\sigma}(x,y)(D^\rho u)(x,y)\overline{(D^\sigma v)(x,y)}dxdy, \tag{2.3}$$

for $u, v \in H^m(\Omega)$.

Due to condition (2.2) and [1, Theorem 5.2], B_Ω satisfies a Gårding inequality of the form

$$B_\Omega[u,u] \geq c\,\|u\|^2_{H^m(\Omega)} - C\,\|u\|^2_{L^2(\Omega)} \text{ for all } u \in H^m(\Omega).$$

Since we are studying a spectral problem, we therefore may assume without loss of generality (introducing a shift by C) that B and B_Ω are H^m-elliptic, i.e., there is a $c > 0$ such that $B[u,u] \geq c\,\|u\|^2_{H^m(\mathbb{R}^d)}$ for all $u \in H^m(\mathbb{R}^d)$ and $B_\Omega[v,v] \geq c\,\|v\|^2_{H^m(\Omega)}$ for all $v \in H^m(\Omega)$, where $\|v\|^2_{H^m(\Omega)} = \sum_{|\alpha|\leq m} \|D^\alpha v\|^2_{L^2(\Omega)}$. (Note that H^m-ellipticity of B_Ω implies H^m-ellipticity for B due to periodicity of the coefficients.)

This allows us to introduce new scalar products on $H^m(\mathbb{R}^d)$ and $H^m(\Omega)$ given by

$$\langle u,v\rangle_{H^m(\mathbb{R}^d)} := B[u,v] \quad \text{and} \quad \langle u,v\rangle_{H^m(\Omega)} := B_\Omega[u,v]$$

which are equivalent to the usual scalar products in $H^m(\mathbb{R}^d)$ and $H^m(\Omega)$, respectively. By $\|\cdot\|_{H^m(\mathbb{R}^d)}$ and $\|\cdot\|_{H^m(\Omega)}$ we denote the associated norms.

[1] This assumption is made for simplicity; in general, we only require d_2 linearly independent vectors in $\mathbb{R}^{d_2}$.

[2] The authors wish to thank Gerd Grubb and Hans-Christoph Grunau for their related remarks.

Definition 2.2. Let $H^{-m}(\mathbb{R}^d)$ denote the dual space of $H^m(\mathbb{R}^d)$. Let $\phi : H^m(\mathbb{R}^d) \to H^{-m}(\mathbb{R}^d)$ be defined by

$$\langle \phi[u], \varphi \rangle = B[u, \varphi] \quad \text{for all} \quad u, \varphi \in H^m(\mathbb{R}^d) \tag{2.4}$$

where the $\langle \cdot, \cdot \rangle$-notation indicates the dual pairing, i.e.,

$$\langle w, \varphi \rangle = w[\overline{\varphi}] \text{ for all } w \in H^{-m}(\mathbb{R}^d),\ \varphi \in H^m(\mathbb{R}^d).$$

ϕ is an isometric isomorphism, and hence the scalar product on $H^{-m}(\mathbb{R}^d)$ given by

$$\langle u, v \rangle_{H^{-m}(\mathbb{R}^d)} := \langle \phi^{-1} u, \phi^{-1} v \rangle_{H^m(\mathbb{R}^d)}$$

induces a norm which coincides with the usual operator sup-norm on $H^{-m}(\mathbb{R}^d)$.

Proposition 2.3. *We define an operator* $\mathrm{L} : D(\mathrm{L}) \to H^{-m}(\mathbb{R}^d)$ *by* $D(\mathrm{L}) := H^m(\mathbb{R}^d) \subset H^{-m}(\mathbb{R}^d)$ *and*

$$\mathrm{L}\, u := \phi u.$$

Then L *is self-adjoint.*

Proof. For $u, v \in H^m(\mathbb{R}^d)$,

$$\begin{aligned}
\langle \mathrm{L}\, u, v \rangle_{H^{-m}(R^d)} &= \langle \phi^{-1}\mathrm{L}\, u, \phi^{-1} v \rangle_{H^m(\mathbb{R}^d)} \\
&= \langle u, \phi^{-1} v \rangle_{H^m(\mathbb{R}^d)} = \overline{\langle \phi^{-1} v, u \rangle_{H^m(\mathbb{R}^d)}} = \overline{B[\phi^{-1} v, u]} \\
&= \overline{\langle v, u \rangle} = \overline{\langle v, u \rangle_{\mathrm{L}^2}} = \langle u, v \rangle_{L^2};
\end{aligned}$$

the last line follows by (2.4). Thus L is symmetric.

Since ϕ is bijective it follows that L is bijective, thus $\mathrm{L}^{-1} : H^{-m}(\mathbb{R}^d) \to H^{-m}(\mathbb{R}^d)$ is defined on the whole space and is also symmetric. Therefore, L^{-1} is self-adjoint. Hence L is self-adjoint. □

3. Floquet transform in $H^m(\mathbb{R}^d)$ and $H^{-m}(\mathbb{R}^d)$

In this section, we recall the Floquet transform on $L^2(\mathbb{R}^d)$ and show that its restriction to $H^m(\mathbb{R}^d)$ is an isometric isomorphism between $H^m(\mathbb{R}^d)$ and a suitable Hilbert space $\mathcal{H}$. By a simple duality argument, we extend the Floquet transform to an isometric isomorphism between $H^{-m}(\mathbb{R}^d)$ and $\mathcal{H}'$.

Definition 3.1. For a lattice $R \subset \mathbb{R}^n$ the reciprocal lattice K consists of those points k in $\mathbb{R}^n$ such that

$$e^{ir\cdot k} = 1$$

for all $r \in R$. The first Brillouin zone associated with a lattice R consists of those points in $\mathbb{R}^n$ whose distance to the origin is smaller than or equal to their distance from any other point in the reciprocal lattice.

The Brillouin zone $\mathcal{K} \subset \mathbb{R}^{d_2}$ for the lattice $\mathbb{Z}^{d_2}$, which corresponds to our periodic cell, is $\mathcal{K} := [-\pi, \pi]^{d_2}$.

Definition 3.2. For all $k \in \mathcal{K}$, we now introduce an extension operator $\mathrm{E}_k : L^2(\Omega) \to L^2_{\mathrm{loc}}(\mathbb{R}^d)$ with

$$(\mathrm{E}_k\, u)(x, y + p) := e^{ik\cdot p} u(x, y)$$

for all $(x, y) \in \Omega$, $p \in \mathbb{Z}^{d_2}$.

The Floquet transform

$$\mathrm{U} : L^2(\mathbb{R}^d) \to L^2(\Omega \times \mathcal{K})$$

is given by

$$(\mathrm{U}\, u)(x, y, k) := \frac{1}{(2\pi)^{d_2/2}} \sum_{n\in\mathbb{Z}^{d_2}} e^{ik\cdot n} u(x, y - n) \quad \text{for } (x, y) \in \Omega, k \in \mathcal{K}.$$

We need the following lemma which together with a proof can be found in [10, Theorem 2.2.5].

Lemma 3.3. U *is an isometric isomorphism and*

$$(\mathrm{U}^{-1}\, v)(x, y) = \frac{1}{(2\pi)^{d_2/2}} \int_{\mathcal{K}} (\mathrm{E}_k\, v(\cdot, \cdot, k))(x, y) dk.$$

The following lemma shows that the formula for U has a canonical extension.

Lemma 3.4. *For all* $u \in L^2(\mathbb{R}^d)$, $k \in \mathcal{K}$ *and* $(x, y) \in \mathbb{R}^d$

$$\mathrm{E}_k[\mathrm{U}\, u(\cdot, \cdot, k)](x, y) = \frac{1}{(2\pi)^{\frac{d_2}{2}}} \sum_{n\in\mathbb{Z}^{d_2}} e^{ik\cdot n} u(x, y - n).$$

Proof. It follows from the definition of E_k that, for $p \in \mathbb{Z}^{d_2}$ and $(x, y) \in \Omega$,

$$\begin{aligned}\mathrm{E}_k[\mathrm{U}\, u(\cdot, \cdot, k)](x, y + p) &= e^{ik\cdot p}\, \mathrm{U}\, u(x, y, k) \\ &= \frac{1}{(2\pi)^{\frac{d_2}{2}}} \sum_{n\in\mathbb{Z}^{d_2}} e^{ik\cdot(n+p)} u(x, y + p - (n + p)) \\ &= \frac{1}{(2\pi)^{\frac{d_2}{2}}} \sum_{\tilde{n}\in\mathbb{Z}^{d_2}} e^{ik\cdot\tilde{n}} u(x, y + p - \tilde{n}).\end{aligned}$$

Noting that $(x, y + p)$ runs through $\mathbb{R}^d$ completes the proof. □

Definition 3.5. For all $k \in \mathcal{K}$, let

$$\mathcal{H}_k := \{u \in H^m(\Omega) : \mathrm{E}_k\, u \in H^m_{\mathrm{loc}}(\mathbb{R}^d)\}.$$

Note that being an element of $\mathcal{H}_k$ requires a weak form of semi-periodic boundary conditions on $\partial\Omega$.

We denote by N_k the mapping

$$N_k : \mathcal{H}_0 \to \mathcal{H}_k, \quad (N_k u)(x, y) := e^{ik\cdot y} u(x, y)$$

and extend it to a mapping $\mathcal{H}_0' \to \mathcal{H}_k'$ by

$$\langle N_k u, \varphi\rangle := \langle u, N_k^{-1}\varphi\rangle \text{ for all } u \in \mathcal{H}_0', \varphi \in \mathcal{H}_k.$$

Let

$$\mathcal{H} = \left\{ u \in L^2(\Omega \times \mathcal{K}) : \forall' k \in \mathcal{K} \quad u(\cdot,\cdot,k) \in \mathcal{H}_k, \right.$$

$$\left. \left\{ \begin{array}{l} \mathcal{K} \to \mathbb{C} \\ k \mapsto \langle N_k^{-1} u(\cdot,\cdot,k), \varphi \rangle_{H^m(\Omega)} \end{array} \right\} \begin{array}{l} \text{measurable for all } \varphi \in \mathcal{H}_0, \\ \text{and } \|u\|_{\mathcal{H}} < \infty \end{array} \right\}$$

where the norm $\|\cdot\|_{\mathcal{H}}$ is induced by the scalar product

$$\langle u, v \rangle_{\mathcal{H}} = \int_{\mathcal{K}} \langle u(\cdot,\cdot,k), v(\cdot,\cdot,k) \rangle_{H^m(\Omega)} dk.$$

$\mathcal{H}$ can be viewed as the space of all functions $u(x,y,k) = (N_k v(k))(x,y)$ with $v \in L^2(\mathcal{K}, \mathcal{H}_0)$. $N_k : \mathcal{H}_0 \to \mathcal{H}_k$ is a homeomorphism, with N_k and N_k^{-1} being uniformly bounded with respect to k in the compact set $\mathcal{K}$, which implies in particular that $\mathcal{H}$ is a Hilbert space.

Lemma 3.6. *Let $M \subseteq \mathbb{R}^{d_2}$ be any open bounded set. Then*

a) *the operator $\mathrm{E}_k : L^2(\Omega) \to L^2(\mathbb{R}^{d_1} \times M)$ is bounded,*

b) *the operator $\mathrm{E}_k : \mathcal{H}_k \to H^m(\mathbb{R}^{d_1} \times M)$ is bounded, and $D^\rho(\mathrm{E}_k u) = \mathrm{E}_k(D^\rho u)$, for $u \in \mathcal{H}_k, |\rho| \leq m$.*

c) *for all $k \in \mathcal{K}$, $\mathcal{H}_k \subseteq H^m(\Omega)$ is closed,*

Proof. a) Let $M \subseteq [-l, l]^{d_2}$. Then

$$\int_{\mathbb{R}^{d_1} \times M} |\, \mathrm{E}_k u \,|^2 \, dxdy \leq (2l)^{d_2} \int_\Omega |\, \mathrm{E}_k u \,|^2 \, dxdy = (2l)^{d_2} \int_\Omega |\, u \,|^2 \, dxdy.$$

b) For all $p \in \mathbb{Z}^{d_2}$, all $\varphi \in C_0^\infty(\Omega + p)$, $u \in \mathcal{H}_k$ and $|\, \rho \,| \leq m$ we have

$$\begin{aligned} \langle D^\rho(\mathrm{E}_k u), \varphi \rangle_{L^2(\mathbb{R}^d)} &= \int_{\mathbb{R}^d} D^\rho(\mathrm{E}_k u) \overline{\varphi} dxdy \\ &= (-1)^{|\rho|} \int_{\mathbb{R}^d} \mathrm{E}_k u D^\rho \overline{\varphi} dxdy \\ &= (-1)^{|\rho|} \int_{\mathbb{R}^d} e^{ik \cdot p} u(x, y-p) D^\rho \overline{\varphi(x,y)} dxdy \\ &= \int_{\mathbb{R}^d} e^{ik \cdot p} (D^\rho u)(x, y-p) \overline{\varphi(x,y)} dxdy \\ &= \int_{\mathbb{R}^d} \mathrm{E}_k(D^\rho u) \overline{\varphi} dxdy \quad = \quad \langle \mathrm{E}_k(D^\rho u), \varphi \rangle_{L^2(\mathbb{R}^d)}. \end{aligned}$$

This implies that $D^\rho(\mathrm{E}_k u) = \mathrm{E}_k(D^\rho u)$. Hence, by part a), for all $|\, \rho \,| \leq m$,

$$\|D^\rho(\mathrm{E}_k u)\|_{L^2(\mathbb{R}^{d_1} \times M)} = \|\mathrm{E}_k(D^\rho u)\|_{L^2(\mathbb{R}^{d_1} \times M)} \leq (2l)^{d_2} \|D^\rho u\|_{L^2(\Omega)}.$$

c) Suppose $(u_\mu) \in \mathcal{H}_k^{\mathbb{N}}$ is a sequence with $u_\mu \to u$ in $H^m(\Omega)$ as $\mu \to \infty$. Part b) proves that $(\mathrm{E}_k u_\mu)$ is a Cauchy sequence in $H^m(\mathbb{R}^{d_1} \times M)$ and hence converges to some $w \in H^m(\mathbb{R}^{d_1} \times M)$. On the other hand, $\mathrm{E}_k u_\mu \to \mathrm{E}_k u$ in $L^2(\mathbb{R}^{d_1} \times M)$ by part a). Hence, $\mathrm{E}_k u = w \in H^m(\mathbb{R}^{d_1} \times M)$, which proves $u \in \mathcal{H}_k$. □

We are now ready to introduce the Floquet transform on $H^m(\mathbb{R}^d)$.

Theorem 3.7. *Let* V *be given by* $\mathrm{V} := \mathrm{U}|_{H^m(\mathbb{R}^d)}$. *For* $u, v \in H^m(\mathbb{R}^d)$ *we have* $\mathrm{V}\,u, \mathrm{V}\,v \in \mathcal{H}$, *and*

$$\int_{\mathcal{K}} B_\Omega[\mathrm{V}\,u(\cdot,\cdot,k), \mathrm{V}\,v(\cdot,\cdot,k)]dk = B[u,v],$$

that is

$$\langle \mathrm{V}\,u, \mathrm{V}\,v\rangle_{\mathcal{H}} = \langle u, v\rangle_{H^m(\mathbb{R}^d)}, \tag{3.1}$$

i.e., $\mathrm{V} : H^m(\mathbb{R}^d) \to \mathcal{H}$ *is an isometry.*

Proof. Let $u \in H^m(\mathbb{R}^d)$ have compact support. Then

$$\mathrm{E}_k[\mathrm{V}\,u(\cdot,\cdot,k)](x,y) = \frac{1}{(2\pi)^{\frac{d_2}{2}}} \sum_{n\in\mathbb{Z}^{d_2}} e^{ik\cdot n} u(x, y-n) \text{ on } \mathbb{R}^d$$

by Lemma 3.4, and hence $(\mathrm{V}\,u)(\cdot,\cdot,k) \in \mathcal{H}_k$ since the sum is locally finite. Furthermore, for $\varphi \in \mathcal{H}_0$,

$$\langle N_k^{-1}(\mathrm{V}\,u)(\cdot,\cdot,k), \varphi\rangle_{H^m(\Omega)} \frac{1}{(2\pi)^{\frac{d_2}{2}}} \sum_{n\in\mathbb{Z}^{d_2}} e^{ik\cdot n} B_\Omega[N_k^{-1}u(\cdot,\cdot-n), \varphi]$$

is a measurable function of k.

With $v \in H^m(\mathbb{R}^d)$ denoting a second compact support function, we get

$$\begin{aligned}
&\langle (\mathrm{V}\,u)(\cdot,\cdot,k), (\mathrm{V}\,v)(\cdot,\cdot,k)\rangle_{H^m(\Omega)} \\
&= \frac{1}{(2\pi)^{d_2}} B_\Omega\Bigg[\sum_{n\in\mathbb{Z}^{d_2}} e^{ik\cdot n} u(\cdot,\cdot-n), \sum_{\widetilde{n}\in\mathbb{Z}^{d_2}} e^{ik\cdot\widetilde{n}} v(\cdot,\cdot-\widetilde{n})\Bigg] \\
&= \frac{1}{(2\pi)^{d_2}} \sum_{n,\widetilde{n}\in\mathbb{Z}^{d_2}} e^{ik\cdot(n-\widetilde{n})} B_\Omega[u(\cdot,\cdot-n), v(\cdot,\cdot-\widetilde{n})].
\end{aligned}$$

Since the sum is finite, this expression is integrable over $\mathcal{K}$ and

$$\int_{\mathcal{K}} \langle (\mathrm{V}\,u)(\cdot,\cdot,k), (\mathrm{V}\,v)(\cdot,\cdot,k)\rangle_{H^m(\Omega)} dk = \sum_{n\in\mathbb{Z}^{d_2}} B_\Omega[u(\cdot,\cdot-n), v(\cdot,\cdot-n)]$$

$$\begin{aligned}
&\sum_{n\in\mathbb{Z}^{d_2}} \sum_{|\rho|,|\sigma|\le m} \int_\Omega a^{\rho\sigma}(x,y) D^\rho u(x,y-n) \overline{D^\sigma v(x,y-n)} dx dy \\
&= \sum_{n\in\mathbb{Z}^{d_2}} \sum_{|\rho|,|\sigma|\le m} \int_{\Omega-(0,n)} a^{\rho\sigma}(x,\widetilde{y}+n) D^\rho u(x,\widetilde{y}) \overline{D^\sigma v(x,\widetilde{y})} dx d\widetilde{y} \\
&= \sum_{|\rho|,|\sigma|\le m} \int_{\mathbb{R}^d} a^{\rho\sigma}(x,y) D^\rho u(x,y) \overline{D^\sigma v(x,y)} dx dy \\
&= B[u,v] \; = \; \langle u,v\rangle_{H^m(\mathbb{R}^d)},
\end{aligned}$$

which shows that $\mathrm{V}\,u, \mathrm{V}\,v \in \mathcal{H}$, and that (3.1) holds.

Now let $u \in H^m(\mathbb{R}^d)$ be arbitrary, and choose a sequence $(u_\mu) \in H^m(\mathbb{R}^d)^{\mathbb{N}}$ of compact support functions converging to u in $H^m(\mathbb{R}^d)$. Since (3.1) holds for compact support functions, $(\mathrm{V}\, u_\mu)$ is a Cauchy sequence in $\mathcal{H}$ and thus converges to some $w \in \mathcal{H}$. On the other hand, by Lemma 3.3, $\mathrm{V}\, u_\mu = \mathrm{U}\, u_\mu \to \mathrm{U}\, u = \mathrm{V}\, u$ in $L^2(\Omega \times \mathcal{K})$. Thus, $\mathrm{V}\, u = w \in \mathcal{H}$, and $\mathrm{V}\, u_\mu \to \mathrm{V}\, u$ in $\mathcal{H}$, whence (3.1) follows for all $u, v \in H^m(\mathbb{R}^d)$. □

We next show that, moreover, V is an isomorphism.

Theorem 3.8. $\mathrm{V} : H^m(\mathbb{R}^d) \to \mathcal{H}$ *is an isometric isomorphism.*

Proof. It remains to show that V is onto. For any $w \in L^2(\Omega \times K)$ we have, by Lemma 3.3,

$$(\mathrm{U}^{-1}\, w)(x, y) = \frac{1}{(2\pi)^{d_2/2}} \int_{\mathcal{K}} (\mathrm{E}_k\, w)(x, y, k)dk.$$

Now let $w \in \mathcal{H}$. Then $D^\alpha w \in L^2(\Omega \times K)$ for $|\alpha| \leq m$. Let $\varphi \in C_0^\infty(\mathbb{R}^d)$, whence

$$\int_{\mathbb{R}^d} (D^\alpha \varphi)(\mathrm{U}^{-1}\, w)dxdy = \frac{1}{(2\pi)^{d_2/2}} \int_{\mathbb{R}^d} D^\alpha \varphi \left(\int_{\mathcal{K}} (\mathrm{E}_k\, w)(x, y, k)dk \right) dxdy.$$

Thus Fubini's theorem, integration by parts, and Lemma 3.6 b) yield

$$\begin{aligned}
\int_{\mathbb{R}^d} (D^\alpha \varphi)(\mathrm{U}^{-1}\, w)dxdy &= \frac{1}{(2\pi)^{d_2/2}} \int_{\mathcal{K}} \left(\int_{\mathbb{R}^d} D^\alpha \varphi(\mathrm{E}_k\, w)(x, y, k)dxdy \right) dk \\
&= \frac{(-1)^{|\alpha|}}{(2\pi)^{d_2/2}} \int_{\mathcal{K}} \left(\int_{\mathbb{R}^d} \varphi D^\alpha (\mathrm{E}_k\, w)(x, y, k)dxdy \right) dk \\
&= \frac{(-1)^{|\alpha|}}{(2\pi)^{d_2/2}} \int_{\mathcal{K}} \left(\int_{\mathbb{R}^d} \varphi\, \mathrm{E}_k (D^\alpha w)(x, y, k)dxdy \right) dk \\
&= \frac{(-1)^{|\alpha|}}{(2\pi)^{d_2/2}} \int_{\mathbb{R}^d} \varphi \left(\int_{\mathcal{K}} \mathrm{E}_k (D^\alpha w)(\cdot, \cdot, k)dk \right) dxdy \\
&= (-1)^{|\alpha|} \int_{\mathbb{R}^d} \varphi\, \mathrm{U}^{-1}(D^\alpha w)dxdy,
\end{aligned}$$

This shows that $\mathrm{U}^{-1}\, w$ is differentiable and $D^\alpha(\mathrm{U}^{-1}\, w) = \mathrm{U}^{-1}(D^\alpha w) \in L^2(\mathbb{R}^d)$. Hence $\mathrm{U}^{-1}\, w \in H^m(\mathbb{R}^d)$ and $\mathrm{V}(\mathrm{U}^{-1}\, w) = \mathrm{U}(\mathrm{U}^{-1}\, w) = w$. □

For the next lemma, in which we extend the Floquet transform to $H^{-m}(\mathbb{R}^d)$, we note that $\mathcal{H}$ is dense in $L^2(\Omega \times \mathcal{K})$ by Lemma 3.3 and the denseness of $H^m(\mathbb{R}^d)$ in $L^2(\mathbb{R}^d)$, together with the fact that U maps $H^m(\mathbb{R}^d)$ into $\mathcal{H}$ by Theorem 3.7.

Lemma 3.9. *The map* $\hat{\mathrm{V}} := (\mathrm{V}^*)^{-1} : H^{-m}(\mathbb{R}^d) \to \mathcal{H}'$ *is an isometric isomorphism and* $\hat{\mathrm{V}}\, |_{L^2(\mathbb{R}^d)} = \mathrm{U}$.

Proof. By Theorem 3.8, $\mathrm{V} : H^m(\mathbb{R}^d) \to \mathcal{H}$ is an isometric isomorphism and thus so is the dual map $V^* : \mathcal{H}' \to H^{-m}(\mathbb{R}^d)$ and hence $\hat{\mathrm{V}}$. For $u \in L^2(\mathbb{R}^d)$, $\varphi \in \mathcal{H}$

we get

$$\begin{aligned}\langle \hat{\mathrm{V}} u, \varphi\rangle &= \langle (\mathrm{V}^*)^{-1} u, \varphi\rangle = \langle (\mathrm{V}^{-1})^* u, \varphi\rangle \\ &= \langle u, \mathrm{V}^{-1}\varphi\rangle = \langle u, \mathrm{V}^{-1}\varphi\rangle_{L^2(\mathbb{R}^d)} = \langle u, \mathrm{U}^{-1}\varphi\rangle_{L^2(\mathbb{R}^d)} \\ &= \langle \mathrm{U}\, u, \varphi\rangle_{L^2(\Omega\times\mathcal{K})},\end{aligned}$$

using Lemma 3.3 in the last line. Thus $\hat{\mathrm{V}} u = \mathrm{U}\, u$. □

4. Floquet-Bloch theory

The aim of this section is to show that the spectrum of the operator L can be determined by considering quasiperiodic operators $\mathrm{L_k}$ on the periodic cell, which establishes the main result of Floquet-Bloch theory also in our general situation of non-smooth coefficients and unbounded periodic cell.

Definition 4.1.

(a) For $k \in \mathcal{K}$, let $D(\mathrm{L_k}) := \mathcal{H}_k$ be the domain of the operator $\mathrm{L_k}$ defined in $\mathcal{H}_k'$ by

$$\mathrm{L_k} : D(\mathrm{L_k}) \subseteq \mathcal{H}_k' \to \mathcal{H}_k', \quad \langle \mathrm{L_k}\, u, \varphi\rangle = B_\Omega[u, \varphi] \text{ for } u, \varphi \in \mathcal{H}_k.$$

Note that $D(L_k)$ is dense in $\mathcal{H}_k'$ since $\mathcal{H}_k$ is dense in $L^2(\Omega)$ and thus, by duality $L^2(\Omega)$ is dense in $\mathcal{H}_k'$.

(b) For many purposes it is useful to consider, as well as $\mathrm{L_k}$, also the operator $\mathfrak{L}_k := N_k^* \,\mathrm{L_k}\, N_k$ in $\mathcal{H}_0'$ with domain $D(\mathfrak{L}_k) := \mathcal{H}_0$. The fact that $D(\mathfrak{L}_k)$ is independent of k is particularly advantageous, e.g., for numerical computations. Since $\mathfrak{L}_k - \lambda = N_k^*(L_k - \lambda)N_k$ and $N_k : \mathcal{H}_0' \to \mathcal{H}_k'$ is homeomorphic, the spectra of $\mathfrak{L}_k$ and $\mathrm{L_k}$ coincide. Thus, clearly, the following theorems also hold with $\sigma(\mathfrak{L}_k)$ in place of $\sigma(\mathrm{L_k})$.

(c) An inner product on $\mathcal{H}_k'$ is given by

$$\langle u, v\rangle_{\mathcal{H}_k'} = \langle \phi_k^{-1} u, \phi_k^{-1} v\rangle_{H^m(\Omega)}$$

with $\phi_k : \mathcal{H}_k \to \mathcal{H}_k'$ denoting the canonical isometric isomorphism, i.e.,

$$\langle \phi_k u, \varphi\rangle := \langle u, \varphi\rangle_{\mathcal{H}_k} = \langle u, \varphi\rangle_{H^m(\Omega)} = B_\Omega[u, \varphi].$$

Lemma 4.2. *The operator* $\mathrm{L_k}$ *is self-adjoint.*

Proof.

$$\begin{aligned}\langle \mathrm{L_k}\, u, v\rangle_{\mathcal{H}_k'} &= \langle \phi_k^{-1}\, \mathrm{L_k}\, u, \phi_k^{-1} v\rangle_{\mathcal{H}_k} = \langle u, \phi_k^{-1} v\rangle_{\mathcal{H}_k} \\ &= \overline{\langle \phi_k^{-1} v, u\rangle_{\mathcal{H}_k}} = \overline{\langle v, u\rangle} = \langle u, v\rangle_{L^2},\end{aligned}$$

whence the operator L_k is symmetric. Moreover, L_k is bijective since ϕ_k is. Hence, L_k^{-1} is symmetric and defined on the whole of $\mathcal{H}_k'$, and therefore self-adjoint. Thus $\mathrm{L_k}$ is self-adjoint. □

We are now in a position to establish the relationship between the spectra of $\mathrm{L_k}$ and L. We first show the following inclusion:

Theorem 4.3. *Let* $\sigma(\mathrm{L})$ *be the spectrum of* L *and* $\sigma(\mathrm{L_k})$ *the spectrum of* $\mathrm{L_k}$. *Then*

$$\sigma(\mathrm{L}) \subseteq \overline{\bigcup_{k\in\mathcal{K}} \sigma(\mathrm{L_k})}$$

Proof. Suppose $\lambda \in \mathbb{R}$, $\lambda \notin \overline{\cup_{k\in\mathcal{K}}\sigma(\mathrm{L_k})}$. Then there is a $\delta > 0$ such that for all $k \in \mathcal{K}$, we have $\operatorname{dist}(\lambda, \sigma(\mathrm{L_k})) \geq \delta$.

We show that for all $f \in H^{-m}(\mathbb{R}^d)$ there exists a unique $u \in H^m(\mathbb{R}^d)$ with $(\mathrm{L}-\lambda)u = f$. For all $k \in \mathcal{K}$, let $w(k) \in \mathcal{H}'_k$ be defined by $\langle w(k), \varphi\rangle = \langle(\phi_{\mathcal{H}}^{-1}\hat{\mathrm{V}}f)(\cdot,\cdot,k), \varphi\rangle_{H^m(\Omega)}$ for $\varphi \in \mathcal{H}_k$, where $\phi_{\mathcal{H}} : \mathcal{H} \to \mathcal{H}'$ is the canonical isometric isomorphism. Then $\|w(k)\|_{\mathcal{H}'_k} = \left\|(\phi_{\mathcal{H}}^{-1}\hat{\mathrm{V}}f)(\cdot,\cdot,k)\right\|_{H^m(\Omega)}$. In order to be able to integrate over $\mathcal{K}$, we need to check measurability with respect to k. Let $z(\cdot,\cdot,k) := N_k^{-1}(\phi_{\mathcal{H}}^{-1}\hat{\mathrm{V}}f)(\cdot,\cdot,k) \in \mathcal{H}_0$. Since $\phi_{\mathcal{H}}^{-1}\hat{\mathrm{V}}f \in \mathcal{H}$, we know that $k \mapsto \langle z(\cdot,\cdot,k), \varphi\rangle_{H^m(\Omega)}$ is measurable for all $\varphi \in \mathcal{H}_0$. Therefore,

$$\left\{ \begin{array}{l} \mathcal{K} \to \mathcal{H}_0 \\ k \mapsto z(\cdot,\cdot,k) \end{array} \right\} \text{ is measurable.} \tag{4.1}$$

Since

$$\|w(k)\|^2_{\mathcal{H}'_k} = \left\|(\phi_{\mathcal{H}}^{-1}\hat{\mathrm{V}}f)(\cdot,\cdot,k)\right\|^2_{H^m(\Omega)} = B_\Omega[N_k(z(\cdot,\cdot,k), N_k(z(\cdot,\cdot,k)],$$

and the r.h.s. contains exponentials and polynomials in k and derivatives up to order m of $z(\cdot,\cdot,k)$, this is measurable with respect to $k \in \mathcal{K}$. Moreover,

$$\int_{\mathcal{K}} \|w(k)\|^2_{\mathcal{H}'_k}\, dk = \left\|\phi_{\mathcal{H}}^{-1}\hat{\mathrm{V}}f\right\|^2_{\mathcal{H}} = \left\|\hat{\mathrm{V}}f\right\|^2_{\mathcal{H}'} = \|f\|^2_{H^{-m}(\mathbb{R}^d)} < \infty. \tag{4.2}$$

Let $v(\cdot,\cdot,k) := (\mathrm{L_k}-\lambda)^{-1}w(k)$ for all $k \in \mathcal{K}$. Then $v(\cdot,\cdot,k) \in \mathcal{H}_k$. As the following Lemma 4.4 shows, $v \in \mathcal{H}$.

Now, let $u := \mathrm{V}^{-1}v \in H^m(\mathbb{R}^d)$. Then for all $\varphi \in H^m(\mathbb{R}^d)$ we get, by Theorem 3.8 and Lemma 3.3

$$\begin{aligned}
\langle(\mathrm{L}-\lambda)u, \varphi\rangle &= B[u,\varphi] - \lambda\langle u,\varphi\rangle_{L^2(\mathbb{R}^d)} \\
&= \int_{\mathcal{K}} \big(B_\Omega[\mathrm{V}\,u(\cdot,\cdot,k), (\mathrm{V}\,\varphi)(\cdot,\cdot,k)] - \lambda\langle(\mathrm{V}\,u)(\cdot,\cdot,k), (\mathrm{V}\,\varphi)(\cdot,\cdot,k)\rangle_{L^2(\Omega)}\big)\, dk \\
&= \int_{\mathcal{K}} \{B_\Omega[v(\cdot,\cdot,k), (\mathrm{V}\,\varphi)(\cdot,\cdot,k)] - \lambda\langle v(\cdot,\cdot,k), (\mathrm{V}\,\varphi)(\cdot,\cdot,k)\rangle_{L^2(\Omega)}\}\, dk \\
&= \int_{\mathcal{K}} \langle(\mathrm{L_k}-\lambda)v(\cdot,\cdot,k), (\mathrm{V}\,\varphi)(\cdot,\cdot,k)\rangle dk \\
&= \int_{\mathcal{K}} \langle w(k), (\mathrm{V}\,\varphi)(\cdot,\cdot,k)\rangle dk = \int_{\mathcal{K}} \langle(\phi_{\mathcal{H}}^{-1}\hat{\mathrm{V}}f)(\cdot,\cdot,k), (\mathrm{V}\,\varphi)(\cdot,\cdot,k)\rangle_{H^m(\Omega)} dk \\
&= \langle\phi_{\mathcal{H}}^{-1}\hat{\mathrm{V}}f, \mathrm{V}\,\varphi\rangle_{\mathcal{H}} = \langle\hat{\mathrm{V}}f, \mathrm{V}\,\varphi\rangle = \langle\mathrm{V}^*\,\hat{\mathrm{V}}f, \varphi\rangle = \langle f, \varphi\rangle.
\end{aligned}$$

Thus $(\mathrm{L}-\lambda)u = f$. So $(\mathrm{L}-\lambda)$ is onto. Since it is self-adjoint, $\mathrm{L}-\lambda$ is also one-to-one. □

Lemma 4.4. *Let λ, δ, $w(k)$, z and v be defined as in the proof of Theorem 4.3. Then $v \in \mathcal{H}$.*

Proof. First, for all $\varphi \in \mathcal{H}_0$,

$$\begin{aligned}\langle N_k^{-1} w(k), \varphi \rangle &= \langle w(k), N_k \varphi \rangle \\ &= \langle (\phi_{\mathcal{H}}^{-1} \hat{\mathrm{V}} f)(\cdot, \cdot, k), N_k \varphi \rangle_{H^m(\Omega)} \\ &= \langle (N_k z)(\cdot, \cdot, k), N_k \varphi \rangle_{H^m(\Omega)} \\ &= B_\Omega [N_k z(\cdot, \cdot, k), N_k \varphi]\end{aligned}$$

is measurable with respect to $k \in \mathcal{K}$ by (4.1). So,

$$\left\{ \begin{array}{l} \mathcal{K} \to \mathcal{H}_0' \\ k \mapsto N_k^{-1} w(k) \end{array} \right\} \text{ is measurable.} \tag{4.3}$$

Next, for $k \in \mathbb{C}^{d_2}$, let $D(\mathfrak{L}_k) := \mathcal{H}_0 \subseteq \mathcal{H}_0'$, $\mathfrak{L}_k : D(\mathfrak{L}_k) \to \mathcal{H}_0'$ be given by

$$\begin{aligned}\langle \mathfrak{L}_k u, \varphi \rangle &:= B_\Omega [N_k u, N_{\overline{k}} \varphi] \\ &= \sum_{|\rho|,\, |\sigma| \le m} \int_\Omega a^{\rho\sigma}(x,y)(D^\rho N_k u)(x,y) \overline{(D^\sigma N_{\overline{k}} \varphi)(x,y)} dx dy\end{aligned}$$

for $u, \varphi \in D(\mathfrak{L}_k) = \mathcal{H}_0$, which for all fixed u and φ depends analytically on $k \in \mathbb{C}^{d_2}$. Clearly, this extends the definition of $\mathfrak{L}_k$ in Definition 4.1 (b) to complex k. By the closedness of $\mathfrak{L}_k$ and [9, Theorem VII.1.3],

$$\left\{ \begin{array}{l} \mathcal{K}_{\mathbb{C}} \to \mathcal{B}(\mathcal{H}_0', \mathcal{H}_0') \\ k \mapsto (\mathfrak{L}_k - \lambda)^{-1} \end{array} \right\} \text{ is analytic} \tag{4.4}$$

on some complex open neighbourhood $\mathcal{K}_{\mathbb{C}}$ of $\mathcal{K}$ which is chosen such that λ is in the resolvent set of $\mathfrak{L}_k$ for all $k \in \mathcal{K}_{\mathbb{C}}$; note that $\operatorname{dist}(\lambda, \sigma(\mathfrak{L}_k)) = \operatorname{dist}(\lambda, \sigma(L_k)) \ge \delta > 0$ for all $k \in \mathcal{K}$.

Even stronger,

$$\left\{ \begin{array}{l} \mathcal{K}_{\mathbb{C}} \to \mathcal{B}(\mathcal{H}_0', \mathcal{H}_0) \\ k \mapsto (\mathfrak{L}_k - \lambda)^{-1} \end{array} \right\} \text{ is analytic} \tag{4.5}$$

which can be seen as follows: (4.4) implies the analyticity of

$$\begin{aligned}k \mapsto \langle (\mathfrak{L}_k - \lambda)^{-1} u, \varphi \rangle &= \langle (\mathfrak{L}_k - \lambda)^{-1} u, \varphi \rangle_{L^2(\Omega)} \\ &= \overline{\langle \varphi, (\mathfrak{L}_k - \lambda)^{-1} u \rangle_{L^2(\Omega)}} \text{ for each } u \in \mathcal{H}_0', \varphi \in \mathcal{H}_0,\end{aligned} \tag{4.6}$$

while (4.5) requires the analyticity of

$$\begin{aligned}k \mapsto \langle (\mathfrak{L}_k - \lambda)^{-1} u, \varphi \rangle_{H^m(\Omega)} &= \overline{\langle \psi, (\mathfrak{L}_k - \lambda)^{-1} u \rangle} \\ &\text{for each } u \in \mathcal{H}_0',\ \varphi \in \mathcal{H}_0,\ \psi = \phi_0 \varphi \in \mathcal{H}_0'.\end{aligned} \tag{4.7}$$

By (4.6), analyticity of the map in (4.7) holds for ψ in the dense subset $\mathcal{H}_0$ of $\mathcal{H}_0'$. Moreover, by Lemma 4.5, the mapping in (4.5) is locally bounded, whence [9, VII.1.1] implies (4.7).

Since $(\mathfrak{L}_k - \lambda)^{-1} = N_k^*(L_k - \lambda)^{-1}N_k = N_k^{-1}(L_k - \lambda)^{-1}N_k$, we get, for each $\varphi \in \mathcal{H}_0$

$$\begin{aligned}\langle N_k^{-1}v(\cdot,\cdot,k), \varphi\rangle_{H^m(\Omega)} &= \langle N_k^{-1}(L_k - \lambda)^{-1}N_k N_k^{-1}w(k), \varphi\rangle_{H^m(\Omega)} \\ &= \langle (\mathfrak{L}_k - \lambda)^{-1}N_k^{-1}w(k), \varphi\rangle_{H^m(\Omega)},\end{aligned}$$

which depends on $k \in \mathcal{K}$ in a measurable way, due to (4.3) and (4.5). Moreover, using Lemma 4.5 again,

$$\begin{aligned}\|v(\cdot,\cdot,k)\|_{\mathcal{H}_k} &\leq C\left\|N_k^{-1}v(\cdot,\cdot,k)\right\|_{\mathcal{H}_0} = C\left\|(\mathfrak{L}_k - \lambda)^{-1}N_k^{-1}w(k)\right\|_{\mathcal{H}_0} \\ &\leq C\left\|N_k^{-1}w(k)\right\|_{\mathcal{H}_0'} \leq C\|w(k)\|_{\mathcal{H}_k'},\end{aligned}$$

because $\|N_k u\|_{H^m(\Omega)} \leq C\|u\|_{H^m(\Omega)}$ for all $u \in H^m(\Omega)$, $k \in \mathcal{K}$. Thus, by (4.2), $v \in \mathcal{H}$. □

Lemma 4.5. *Let $\lambda \in \mathbb{R}$ be such that $dist(\lambda, \sigma(\mathfrak{L}_k)) \geq \delta > 0$ for all $k \in \mathcal{K}$. Then, for some constant $C = C(\lambda)$,*

$$\left\|(\mathfrak{L}_k - \lambda)^{-1}g\right\|_{\mathcal{H}_0} \leq C\|g\|_{\mathcal{H}_0'} \quad \textit{for} \quad \textit{all}\, g \in \mathcal{H}_0', k \in \mathcal{K}.$$

Proof. Due to the condition on λ,

$$\left\|(\mathfrak{L}_k - \lambda)^{-1}g\right\|_{\mathcal{H}_0'} \leq \frac{1}{\delta}\|g\|_{\mathcal{H}_0'} \quad (g \in \mathcal{H}_0'). \tag{4.8}$$

Moreover, for $u \in \mathcal{H}_0$,

$$\begin{aligned}\langle(\mathfrak{L}_k - \lambda)u, u\rangle &= \langle \mathrm{L_k}\, N_k u, N_k u\rangle - \lambda\langle u, u\rangle \\ &= B_\Omega[N_k u, N_k u] - \lambda\langle u, u\rangle \\ &\geq \|N_k u\|^2_{\mathcal{H}_k} - |\lambda|\,\|u\|_{\mathcal{H}_0'}\|u\|_{\mathcal{H}_0} \\ &\geq c\|u\|^2_{\mathcal{H}_0} - |\lambda|\,\|u\|_{\mathcal{H}_0'}\|u\|_{\mathcal{H}_0}\end{aligned}$$

and $\langle(\mathfrak{L}_k - \lambda)u, u\rangle \leq \|(\mathfrak{L}_k - \lambda)u\|_{\mathcal{H}_0'}\|u\|_{\mathcal{H}_0}$, implying

$$\|u\|_{\mathcal{H}_0} \leq C(\|u\|_{\mathcal{H}_0'} + \|(\mathfrak{L}_k - \lambda)u\|_{\mathcal{H}_0'}),$$

which gives the assertion, using (4.8) and $u := (\mathfrak{L}_k - \lambda)^{-1}g$. □

As a preparation for the reverse spectral inclusion we prove

Lemma 4.6. *Let $\psi \in C_0^\infty(\mathbb{R}^{d_2})$ be real-valued, $u \in H^m_{\mathrm{loc}}(\mathbb{R}^d)$, and suppose that $\hat{u}$ given by $\hat{u}(x,y) := \psi(y)u(x,y)$ is in $H^m(\mathbb{R}^d)$. Then for all $v \in H^m(\mathbb{R}^d)$*

$$B[\hat{u}, v] = \sum_{|\rho|,|\sigma|\leq m}\int_{\mathbb{R}^d} a^{\rho\sigma}D^\rho u\overline{D^\sigma(\psi v)}dx + \sum_{|\rho|,|\sigma|\leq m}\int_{\mathbb{R}^d} b^{\rho\sigma}D^\rho u\overline{D^\sigma v}dx$$

where $b^{\rho\sigma} = \sum_{\beta\neq 0,|\beta|\leq m} c^{\rho\sigma\beta}D^\beta\psi$ with some $c^{\rho\sigma\beta} \in L^\infty(\mathbb{R}^d)$.

Proof.

$$\begin{aligned}
B[\hat{u},v] &= \sum_{|\alpha|,|\sigma|\le m} \int_{\mathbb{R}^d} a^{\alpha\sigma} D^\alpha \hat{u}\overline{(D^\sigma v)}dx \\
&= \sum_{|\alpha|,|\sigma|\le m} \sum_{\rho\le\alpha} \binom{\alpha}{\rho} \int_{\mathbb{R}^d} a^{\alpha\sigma}(D^{\alpha-\rho}\psi)(D^\rho u)\overline{(D^\sigma v)}dx \\
&= \sum_{|\rho|,|\sigma|\le m} \sum_{|\alpha|\le m,\alpha\ge\rho} \binom{\alpha}{\rho} \int_{\mathbb{R}^d} a^{\alpha\sigma}(D^{\alpha-\rho}\psi)(D^\rho u)\overline{(D^\sigma v)}dx.
\end{aligned}$$

Also

$$\begin{aligned}
&\sum_{|\alpha|,|\rho|\le m} \int_{\mathbb{R}^d} a^{\rho\alpha}(D^\rho u)\overline{D^\alpha(\psi v)}dx \\
&\quad= \sum_{|\alpha|,|\rho|\le m} \sum_{\sigma\le\alpha} \binom{\alpha}{\sigma} \int_{\mathbb{R}^d} a^{\rho\alpha}(D^\rho u)(D^{\alpha-\sigma}\psi)\overline{(D^\sigma v)}dx \\
&\quad= \sum_{|\rho|,|\sigma|\le m} \sum_{|\alpha|\le m,\alpha\ge\sigma} \binom{\alpha}{\sigma} \int_{\mathbb{R}^d} a^{\rho\alpha}(D^\rho u)(D^{\alpha-\sigma}\psi)\overline{(D^\sigma v)}dx.
\end{aligned}$$

Thus

$$B[\hat{u},v] = \sum_{|\rho|,|\sigma|\le m} \int_{\mathbb{R}^d} a^{\rho\sigma}(D^\rho u)\overline{D^\sigma(\psi v)}dx + \sum_{|\rho|,|\sigma|\le m} \int_{\mathbb{R}^d} b^{\rho\sigma} D^\rho u \overline{D^\sigma v}dx$$

where

$$\begin{aligned}
b^{\rho\sigma} &= \sum_{|\alpha|\le m,\alpha\ge\rho} \binom{\alpha}{\rho} a^{\alpha\sigma}(D^{\alpha-\rho}\psi) - \sum_{|\alpha|\le m,\alpha\ge\sigma} \binom{\alpha}{\sigma} a^{\rho\alpha}(D^{\alpha-\sigma}\psi) \\
&= \sum_{|\beta|\le m} c^{\rho\sigma\beta} D^\beta\psi.
\end{aligned}$$

We have $c^{\rho\sigma 0} = 0$ and $c^{\rho\sigma\beta} \in L^\infty(\mathbb{R}^d)$. □

Theorem 4.7.

$$\sigma(\mathrm{L}) \supseteq \overline{\bigcup_{k\in\mathcal{K}} \sigma(\mathrm{L_k})}$$

Proof. It suffices to prove this inclusion without the closure sign since $\sigma(L)$ is closed. Let $\lambda \in \sigma(\mathrm{L_k})$ for some $k \in \mathcal{K}$. Then there is a singular sequence $(u_n) \in D(\mathrm{L_k})^{\mathbb{N}}$ such that $\|u_n\|_{\mathcal{H}'_k} = 1$ and $\|(\mathrm{L_k} - \lambda)u_n\|_{\mathcal{H}'_k} \to 0$ as $n \to \infty$.

Let $\eta \in C^\infty(\mathbb{R})$ take values in the interval $[0,1]$ such that

$$\eta(y) = \begin{cases} 1, & y \le 1 \\ 0, & y \ge 2 \end{cases}.$$

Further, for $l \ge 1$, let $\eta_l \in C^\infty(\mathbb{R}^d)$ be given by

$$\eta_l(x,y) := \eta(|y_1|/l)\eta(|y_2|/l)\cdots\eta(|y_{d_2}|/l).$$

We define

$$u_{n,l}(x,y) := \eta_l(x,y)(\mathrm{E}_k\, u_n)(x,y).$$

Then $u_{n,l} \in H^m(\mathbb{R}^d)$. Our aim is to show that, for a suitable choice of $l = l(n)$, $u_{n,l}$ gives a singular sequence for L.

We introduce the function $\zeta_l(x,y) = \eta\left(\frac{|y_1|}{\sqrt{l}} - \sqrt{l}+1\right)\cdots\eta\left(\frac{|y_{d_2}|}{\sqrt{l}} - \sqrt{l}+1\right)$. Then $|(D^\alpha\zeta_l)(x,y)| \leq C$ for some constant C and for all $|\alpha| \leq m$ and all $l \geq 1$, and moreover, $\frac{|y_i|}{\sqrt{l}} - \sqrt{l}+1 \leq 1$ if $|y_i| \leq l$ and $\frac{|y_i|}{\sqrt{l}} - \sqrt{l}+1 \geq 2$ if $|y_i| \geq l+\sqrt{l}$.

Now,

$$\begin{aligned}
\|u_{n,l}\|_{H^{-m}(\mathbb{R}^d)} &= \sup_{\varphi\in H^m(\mathbb{R}^d),\varphi\neq 0} \frac{|\langle u_{n,l},\varphi\rangle|}{\|\varphi\|_{H^m(\mathbb{R}^d)}} \\
&= \sup_{\varphi\in H^m(\mathbb{R}^d),\varphi\neq 0} \frac{\left|\langle u_{n,l},\varphi\rangle_{L^2(\mathbb{R}^d)}\right|}{\|\varphi\|_{H^m(\mathbb{R}^d)}} \\
&\geq \sup_{\psi\in\mathcal{H}_k,\psi\neq 0} \frac{\left|\langle u_{n,l},\zeta_l(\mathrm{E}_k\,\psi)\rangle_{L^2(\mathbb{R}^d)}\right|}{\|\zeta_l(\mathrm{E}_k\,\psi)\|_{H^m(\mathbb{R}^d)}}.
\end{aligned} \tag{4.9}$$

Moreover, we obtain for $l \geq 1$,

$$\begin{aligned}
\|\zeta_l(\mathrm{E}_k\,\psi)\|^2_{H^m(\mathbb{R}^d)} &\leq C\sum_{|\alpha|\leq m} \|D^\alpha(\zeta_l\,\mathrm{E}_k\,\psi)\|^2_{L^2(\mathbb{R}^d)} \\
&\leq C\sum_{|\alpha|\leq m}\sum_{\beta\leq\alpha}\begin{pmatrix}\alpha\\ \beta\end{pmatrix}\left\|(D^{\alpha-\beta}\zeta_l)(D^\beta(\mathrm{E}_k\,\psi))\right\|^2_{L^2(\mathbb{R}^d)} \\
&\leq C\sum_{|\alpha|\leq m}\sum_{\beta\leq\alpha}\left\|D^\beta(\mathrm{E}_k\,\psi)\right\|^2_{L^2(\mathbb{R}^{d_1}\times[-l-\sqrt{l},l+\sqrt{l}]^{d_2})},
\end{aligned}$$

and thus by Lemma 3.6 b) we have

$$\begin{aligned}
\|\zeta_l(\mathrm{E}_k\,\psi)\|^2_{H^m(\mathbb{R}^d)} &\leq C\sum_{|\alpha|\leq m}\sum_{\beta\leq\alpha}\left\|\mathrm{E}_k(D^\beta\psi)\right\|^2_{L^2(\mathbb{R}^{d_1}\times[-l-\sqrt{l},l+\sqrt{l}]^{d_2})} \\
&\leq C\sum_{|\alpha|\leq m}\sum_{\beta\leq\alpha} l^{d_2}\left\|D^\beta\psi\right\|^2_{L^2(\Omega)} \\
&\leq Cl^{d_2}\,\|\psi\|^2_{H^m(\Omega)}\,.
\end{aligned} \tag{4.10}$$

Furthermore

$$\begin{aligned}
|\langle u_{n,l},\zeta_l(\mathrm{E}_k\,\psi)\rangle_{L^2(\mathbb{R}^d)}| &= \left|\int_{\mathbb{R}^d}\eta_l\zeta_l(\mathrm{E}_k\,u_n)\overline{(\mathrm{E}_k\,\psi)}dxdy\right| \\
&\geq \left|\int_{\mathbb{R}^{d_1}\times[-l,l]^{d_2}}(\mathrm{E}_k\,u_n)\overline{(\mathrm{E}_k\,\psi)}dxdy\right| \\
&\quad - \int_{\mathbb{R}^{d_1}\times([-l-\sqrt{l},l+\sqrt{l}]^{d_2}\setminus[-l,l]^{d_2})}|\,\mathrm{E}_k\,u_n||\,\mathrm{E}_k\,\psi|dxdy \\
&=: A - B,
\end{aligned}$$

with

$$A = (2l)^{d_2} \left| \int_\Omega u_n \overline{\psi} dx dy \right|, \quad B \le \mathrm{vol}(W_l) \int_\Omega |u_n||\psi| dx dy$$

where $W_l = [-l - \sqrt{l}, l + \sqrt{l}]^{d_2} \setminus [-l, l]^{d_2}$. We may estimate the volume of W_l in the following way:

$$\begin{aligned}
\mathrm{vol}(W_l) &= \mathrm{vol}[-l - \sqrt{l}, l + \sqrt{l}]^{d_2} - \mathrm{vol}[-l, l]^{d_2} \\
&= (2(l + \sqrt{l}))^{d_2} - (2l)^{d_2} \\
&= 2^{d_2} \left(\sum_{\nu=0}^{d_2} \begin{pmatrix} d_2 \\ \nu \end{pmatrix} l^\nu \sqrt{l}^{d_2 - \nu} - l^{d_2} \right) \\
&= 2^{d_2} \left(\sum_{\nu=0}^{d_2 - 1} \begin{pmatrix} d_2 \\ \nu \end{pmatrix} l^{\frac{\nu + d_2}{2}} \right) \le C l^{d_2 - \frac{1}{2}}.
\end{aligned}$$

From this it follows that

$$B \le C l^{d_2 - \frac{1}{2}} \|u_n\|_{L^2(\Omega)} \|\psi\|_{L^2(\Omega)} \le C l^{d_2 - \frac{1}{2}} \|u_n\|_{L^2(\Omega)} \|\psi\|_{H^m(\Omega)},$$

yielding

$$\begin{aligned}
|\langle u_{n,l}, \zeta_l(\mathrm{E}_k \psi)\rangle_{L^2(\mathbb{R}^d)}| &\ge A - B \\
&\ge (2l)^{d_2} \left(\left| \int_\Omega u_n \overline{\psi} dx dy \right| - \frac{C}{\sqrt{l}} \|u_n\|_{L^2(\Omega)} \|\psi\|_{H^m(\Omega)} \right)
\end{aligned}$$

and therefore, using (4.10),

$$\frac{|\langle u_{n,l}, \zeta_l \, \mathrm{E}_k \, \psi\rangle_{L^2(\mathbb{R}^d)}|}{\|\zeta_l(\mathrm{E}_k \, \psi)\|_{H^m(\mathbb{R}^d)}} \ge c l^{\frac{d_2}{2}} \left(\frac{|\langle u_n, \psi\rangle_{L^2(\Omega)}|}{\|\psi\|_{H^m(\Omega)}} - \frac{C}{\sqrt{l}} \|u_n\|_{L^2(\Omega)} \right). \tag{4.11}$$

Thus from (4.9) and (4.11) it follows that

$$\|u_{n,l}\|_{H^{-m}(\mathbb{R}^d)} \ge c l^{\frac{d_2}{2}} \left(\|u_n\|_{\mathcal{H}_k'} - \frac{C}{\sqrt{l}} \|u_n\|_{L^2(\Omega)} \right).$$

Note that $\|u_n\|_{\mathcal{H}_k'} = 1$. By choosing $l = l(n)$ large enough so that $\frac{C}{\sqrt{l}} \|u_n\|_{L^2(\Omega)} \le \frac{1}{2}$ we then obtain

$$\|u_{n,l}\|_{H^{-m}(\mathbb{R}^d)} \ge c l^{\frac{d_2}{2}}. \tag{4.12}$$

Now let $\varphi \in H^m(\mathbb{R}^d)$. By Lemma 4.6 we see

$$\begin{aligned}
\langle (\mathrm{L} - \lambda) u_{n,l}, \varphi \rangle &= B[u_{n,l}, \varphi] - \lambda \langle u_{n,l}, \varphi \rangle_{L^2(\mathbb{R}^d)} \\
&= \sum_{|\rho|, |\sigma| \le m} \int_{\mathbb{R}^d} a^{\rho\sigma} D^\rho(\mathrm{E}_k \, u_n) \overline{D^\sigma(\eta_l \varphi)} dx dy - \lambda \int_{\mathbb{R}^d} \eta_l(\mathrm{E}_k \, u_n) \overline{\varphi} dx dy \\
&\quad + \sum_{|\rho|, |\sigma| \le m} \int_{\mathbb{R}^d} b^{\rho\sigma} D^\rho(\mathrm{E}_k \, u_n) \overline{D^\sigma \varphi} dx dy
\end{aligned} \tag{4.13}$$

where

$$b^{\rho\sigma} = \sum_{\beta\neq 0, |\beta|\leq m} c^{\rho\sigma\beta} D^{\beta}\eta_l,$$

whence $| b^{\rho\sigma} |\leq C/l$ for $l \geq 1$, and $b^{\rho\sigma} = 0$ outside $\mathbb{R}^{d_1} \times [-2l, 2l]^{d_2}$.

Hence, using Lemma 3.6 b),

$$\begin{aligned}
&\left| \sum_{|\rho|,|\sigma|\leq m} \int_{\mathbb{R}^d} b^{\rho\sigma} D^{\rho}(\mathrm{E}_k\, u_n)\overline{D^{\sigma}\varphi} dxdy \right| \\
&\quad \leq \sum_{|\rho|,|\sigma|\leq m} \frac{C}{l} \left\| \mathrm{E}_k(D^{\rho}u_n) \right\|_{L^2(\mathbb{R}^{d_1}\times[-2l,2l]^{d_2})} \left\| D^{\sigma}\varphi \right\|_{L^2(\mathbb{R}^d)} \\
&\quad \leq \sum_{|\rho|,|\sigma|\leq m} \frac{C}{l} l^{d_2/2} \left\| D^{\rho}u_n \right\|_{L^2(\Omega)} \left\| D^{\sigma}\varphi \right\|_{L^2(\mathbb{R}^d)} \\
&\quad \leq C l^{d_2/2-1} \left\| u_n \right\|_{H^m(\Omega)} \left\| \varphi \right\|_{H^m(\mathbb{R}^d)} .
\end{aligned} \tag{4.14}$$

For the first two terms on the right-hand side of equation (4.13), we have

$$\begin{aligned}
&\sum_{|\rho|,|\sigma|\leq m} \int_{\mathbb{R}^d} a^{\rho\sigma} D^{\rho}(\mathrm{E}_k\, u_n)\overline{D^{\sigma}(\eta_l\varphi)} dxdy - \lambda \int_{\mathbb{R}^d} (\mathrm{E}_k\, u_n)\overline{\eta_l\varphi} dxdy \\
&= \sum_{p\in\mathbb{Z}^{d_2}} \int_{\Omega+(0,p)} \left[\sum_{|\rho|,|\sigma|\leq m} a^{\rho\sigma} D^{\rho}(\mathrm{E}_k\, u_n)\overline{D^{\sigma}(\eta_l\varphi)} - \lambda(\mathrm{E}_k\, u_n)\overline{\eta_l\varphi} \right] dxdy \\
&= \sum_{p\in\mathbb{Z}^{d_2}} \int_{\Omega} \sum_{|\rho|,|\sigma|\leq m} a^{\rho\sigma}(x,y) D^{\rho}(\mathrm{E}_k\, u_n)(x,y+p)\overline{D^{\sigma}(\eta_l\varphi)}(x,y+p) dxdy \\
&\qquad - \sum_{p\in\mathbb{Z}^{d_2}} \int_{\Omega} \lambda(\mathrm{E}_k\, u_n)(x,y+p)\overline{\eta_l\varphi}(x,y+p) dxdy \\
&= \sum_{p\in\mathbb{Z}^{d_2}} \Big(B_{\Omega}[(\mathrm{E}_k\, u_n)(\cdot,\cdot+p), (\eta_l\varphi)(\cdot,\cdot+p)] \\
&\qquad - \lambda\langle(\mathrm{E}_k\, u_n)(\cdot,\cdot+p), (\eta_l\varphi)(\cdot,\cdot+p)\rangle_{L^2(\Omega)} \Big) \\
&= \sum_{p\in\mathbb{Z}^{d_2}} \left(B_{\Omega}[e^{ik\cdot p}u_n, (\eta_l\varphi)(\cdot,\cdot+p)] - \lambda\langle e^{ik\cdot p}u_n, (\eta_l\varphi)(\cdot,\cdot+p)\rangle_{L^2(\Omega)} \right) \\
&= \sum_{p\in\mathbb{Z}^{d_2}} \langle(\mathrm{L}_\mathrm{k} - \lambda)u_n, e^{-ik\cdot p}(\eta_l\varphi)(\cdot,\cdot+p)\rangle \\
&= \langle(\mathrm{L}_\mathrm{k} - \lambda)u_n, \sum_{p\in\mathbb{Z}^{d_2}} e^{-ik\cdot p}(\eta_l\varphi)(\cdot,\cdot+p)\rangle \\
&= (2\pi)^{d_2/2}\langle(\mathrm{L}_\mathrm{k} - \lambda)u_n, \mathrm{U}(\eta_l\varphi)(\cdot,\cdot,k)\rangle \\
&= (2\pi)^{d_2/2}\langle(\mathrm{L}_\mathrm{k} - \lambda)u_n, \mathrm{V}(\eta_l\varphi)(\cdot,\cdot,k)\rangle.
\end{aligned}$$

Noting that $\mathrm{V}(\eta_l\varphi)(\cdot,\cdot,k) \in \mathcal{H}_k$ by Theorem 3.8, we get from (4.13) and (4.14)

$$\begin{aligned} |\langle(\mathrm{L}-\lambda)u_{n,l},\varphi\rangle| &\leq C\,\|(\mathrm{L_k}-\lambda)u_n\|_{\mathcal{H}'_k}\,\|\mathrm{V}(\eta_l\varphi)(\cdot,\cdot,k)\|_{\mathcal{H}_k} \\ &\quad + Cl^{d_2/2-1}\,\|u_n\|_{H^m(\Omega)}\,\|\varphi\|_{H^m(\mathbb{R}^d)}\,. \end{aligned} \tag{4.15}$$

Now,

$$\begin{aligned} \|\mathrm{V}(\eta_l\varphi)(\cdot,\cdot,k)\|_{\mathcal{H}_k} &= (2\pi)^{-d_2/2}\left\|\sum_{p\in\mathbb{Z}^{d_2}} e^{ik\cdot p}(\eta_l\varphi)(\cdot,\cdot-p)\right\|_{H^m(\Omega)} \\ &= (2\pi)^{-d_2/2}\left\|\sum_{\substack{p\in[-2l,2l+1]^{d_2},\\ p\in\mathbb{Z}^{d_2}}} e^{ik\cdot p}(\eta_l\varphi)(\cdot,\cdot-p)\right\|_{H^m(\Omega)} \\ &\leq (2\pi)^{-d_2/2}\sum_{\substack{p\in[-2l,2l+1]^{d_2},\\ p\in\mathbb{Z}^{d_2}}} \|(\eta_l\varphi)(\cdot,\cdot-p)\|_{H^m(\Omega)} \\ &\leq C\sum_{\substack{p\in[-2l,2l+1]^{d_2},\\ p\in\mathbb{Z}^{d_2}}} \|\varphi(\cdot,\cdot-p)\|_{H^m(\Omega)} \\ &\leq Cl^{d_2/2}\left(\sum_{\substack{p\in[-2l,2l+1]^{d_2},\\ p\in\mathbb{Z}^{d_2}}} \|\varphi(\cdot,\cdot-p)\|^2_{H^m(\Omega)}\right)^{\frac{1}{2}} \\ &\leq Cl^{d_2/2}\,\|\varphi\|_{H^m(\mathbb{R}^d)}\,. \end{aligned}$$

Thus, by (4.15),

$$\frac{|\langle(\mathrm{L}-\lambda)u_{n,l},\varphi\rangle|}{\|\varphi\|_{H^m(\mathbb{R}^d)}} \leq C\,\|(\mathrm{L_k}-\lambda)u_n\|_{\mathcal{H}'_k}\,l^{d_2/2} + Cl^{d_2/2-1}\,\|u_n\|_{H^m(\Omega)}$$

giving

$$\|(\mathrm{L}-\lambda)u_{n,l}\|_{H^{-m}(\mathbb{R}^d)} \leq Cl^{d_2/2}\left(\|(\mathrm{L_k}-\lambda)u_n\|_{\mathcal{H}'_k} + \frac{1}{l}\,\|u_n\|_{H^m(\Omega)}\right)$$

and therefore, using (4.12),

$$\frac{\|(\mathrm{L}-\lambda)u_{n,l}\|_{H^{-m}(\mathbb{R}^d)}}{\|u_{n,l}\|_{H^{-m}(\mathbb{R}^d)}} \leq C\left(\|(\mathrm{L_k}-\lambda)u_n\|_{\mathcal{H}'_k} + \frac{1}{l}\,\|u_n\|_{H^m(\Omega)}\right).$$

Now for each n, choose $l_n := l(n)$ such that

$$\frac{\|u_n\|_{H^m(\Omega)}}{l_n} \leq \frac{1}{n} \qquad \text{as well as} \qquad \frac{C}{\sqrt{l}}\,\|u_n\|_{L^2(\Omega)} \leq \frac{1}{2}$$

which we required earlier in the proof. Then

$$\frac{\|(\mathrm{L}-\lambda)u_{n,l}\|_{H^{-m}(\mathbb{R}^d)}}{\|u_{n,l}\|_{H^{-m}(\mathbb{R}^d)}} \leq C\left(\|(\mathrm{L}_k-\lambda)u_n\|_{\mathcal{H}'_k}+\frac{1}{n}\right) \to 0$$

as $n \to \infty$. Hence, $\lambda \in \sigma(L)$. □

5. Equality of the H^{-m} and L^2-spectra

In the final section we show that the spectrum of the operator L in $H^{-m}(\mathbb{R}^d)$ coincides with the spectrum of the canonical self-adjoint operator in $L^2(\mathbb{R}^d)$ generated by the bilinear form B. We do this on an abstract Hilbert space level.

Suppose $B : D \times D \to \mathbb{C}$ is a closed Hermitian sesquilinear form and $D \subseteq H$ is dense in the Hilbert space H. E.g., for the form B defined in (2.1), we choose $H := L^2(\mathbb{R}^d)$ and $D := H^m(\mathbb{R}^d)$. Suppose also $\|u\|_D^2 := B[u,u] \geq c\,\|u\|_H^2$ for some $c > 0$. Then $(D, B[\cdot,\cdot])$ is a Hilbert space.

Let $\mathrm{L} : D \subseteq D' \to D'$, with $\langle \mathrm{L}\,u, \varphi\rangle = B[u,\varphi]$, and $\widetilde{\mathrm{L}} : D(\widetilde{\mathrm{L}}) \subseteq H \to H$ defined by $\langle \widetilde{\mathrm{L}}\,u, \varphi\rangle_H = B[u,\varphi]$ for all $\varphi \in D$, with domain

$$D(\widetilde{\mathrm{L}}) := \{u \in D : D \ni \varphi \to B[u,\varphi] \text{ is bounded in } H\}.$$

$D(\tilde{L})$ is dense in H, see [14]. Thus, L corresponds to the operator studied in the previous sections, and $\widetilde{\mathrm{L}}$ is the standard selfadjoint operator in H associated with B. Note that $D(\widetilde{\mathrm{L}})$ is not easy to handle directly if B is the form (2.1) with discontinuous coefficients $a^{\rho\sigma}$. For example, multiplication with C_0^∞-functions may lead out of $D(\widetilde{\mathrm{L}})$, which causes severe problems when one tries to prove an analogue of Theorem 4.7 for $\widetilde{\mathrm{L}}$ directly.

L and $\widetilde{\mathrm{L}}$ are self-adjoint in D' and H, respectively. We intend to show $\sigma(\mathrm{L}) = \sigma(\widetilde{\mathrm{L}})$.

Lemma 5.1. *Let $\lambda \notin \sigma(\widetilde{\mathrm{L}})$. Then for all $u \in D(\widetilde{\mathrm{L}})$*

$$\|u\|_D \leq C_\lambda \left\|(\widetilde{\mathrm{L}}-\lambda)u\right\|_H.$$

Proof. Since $\lambda \in \rho(\widetilde{\mathrm{L}})$,

$$\|u\|_H \leq \widetilde{C}_\lambda \left\|(\widetilde{\mathrm{L}}-\lambda)u\right\|_H$$

for all $u \in D(\widetilde{\mathrm{L}})$. Now, $\langle(\widetilde{\mathrm{L}}-\lambda)u,u\rangle_H = B[u,u] - \lambda\langle u,u\rangle_H$, giving

$$\|u\|_D^2 \leq \lambda\,\|u\|_H^2 + \|u\|_H \left\|(\widetilde{\mathrm{L}}-\lambda)u\right\|_H \leq C_\lambda^2 \left\|(\widetilde{\mathrm{L}}-\lambda)u\right\|_H^2.$$

□

Lemma 5.2. *For all $u \in D$ and for all $\psi \in D(\widetilde{\mathrm{L}})$*

$$\langle Lu, \psi\rangle = \langle u, \widetilde{\mathrm{L}}\,\psi\rangle_H.$$

Proof. $\langle u, \widetilde{\mathrm{L}}\,\psi\rangle_H = \overline{\langle \widetilde{\mathrm{L}}\,\psi, u\rangle_H} = \overline{B[\psi,u]} = B[u,\psi] = \langle \mathrm{L}\,u, \psi\rangle$. □

Theorem 5.3. *The spectra of* L *and* $\widetilde{\mathrm{L}}$ *are the same: i.e.,* $\sigma(\mathrm{L}) = \sigma(\widetilde{\mathrm{L}})$.

Proof. As both operators are self-adjoint, it suffices to consider real λ. First let $\lambda \in \mathbb{R} \cap \rho(\mathrm{L})$. Then for all $f \in D'$ there is a $u \in D$ such that $(\mathrm{L} - \lambda)u = f$. This implies in particular

$$\forall f \in H \; \exists u \in D \; \forall \varphi \in D \; \langle (\mathrm{L} - \lambda)u, \varphi \rangle = \langle f, \varphi \rangle_H$$

which, since $\langle (\mathrm{L} - \lambda)u, \varphi \rangle = B[u, \varphi] - \lambda \langle u, \varphi \rangle_H$, shows that $\varphi \mapsto B[u, \varphi]$ is a bounded linear functional with respect to the H-norm. Thus $u \in D(\widetilde{\mathrm{L}})$ and therefore $(\widetilde{\mathrm{L}} - \lambda)u = f$. Hence $\widetilde{\mathrm{L}} - \lambda$ is onto. As $\widetilde{\mathrm{L}} - \lambda$ is self-adjoint, we conclude $\lambda \in \rho(\widetilde{\mathrm{L}})$.

Next assume that $\lambda \in \rho(\widetilde{\mathrm{L}})$. Then if $(\mathrm{L} - \lambda)u = 0$ for some $u \in D$, we get as above that $u \in D(\widetilde{\mathrm{L}})$ and $(\widetilde{\mathrm{L}} - \lambda)u = 0$, so $u = 0$. Hence $(\mathrm{L} - \lambda)$ is one-to-one. As L is self-adjoint we therefore get that the range of $(\mathrm{L} - \lambda)$ is dense in D'. For all $\psi \in D$, $u \in D$ we have (letting $\varphi = (\widetilde{\mathrm{L}} - \lambda)^{-1}\psi$)

$$\frac{\langle u, \psi \rangle_H}{\|\psi\|_H} = \frac{\langle u, (\widetilde{\mathrm{L}} - \lambda)\varphi \rangle_H}{\left\| (\widetilde{\mathrm{L}} - \lambda)\varphi \right\|_H} = \frac{\langle (\mathrm{L} - \lambda)u, \varphi \rangle_H}{\left\| (\widetilde{\mathrm{L}} - \lambda)\varphi \right\|_H}$$

the last equality following by Lemma 5.2. So by Lemma 5.1,

$$\frac{|\, \langle u, \psi \rangle_H \,|}{\|\psi\|_H} \leq C_\lambda \frac{\|(\mathrm{L} - \lambda)u\|_{D'} \, \|\varphi\|_D}{\|\varphi\|_D} = C_\lambda \, \|(\mathrm{L} - \lambda)u\|_{D'} \, .$$

Hence, $\|u\|_H \leq C_\lambda \, \|(\mathrm{L} - \lambda)u\|_{D'}$ for all $u \in D$ and

$$\|u\|_{D'} \leq C \, \|u\|_H \leq C_\lambda \, \|(\mathrm{L} - \lambda)u\|_{D'} \, .$$

Thus $(\mathrm{L} - \lambda)^{-1}$ is bounded, and so $\lambda \in \rho(\mathrm{L})$. □

References

[1] R.A. Adams, J.J.F. Fournier: *Sobolev Spaces, Second edition*, Academic Press, New York, 2003.

[2] J.M. Combes, B. Gralak, and A. Tip. Spectral properties of absorptive photonic crystals. In *Waves in periodic and random media* (*South Hadley, MA,* 2002), volume 339 of *Contemp. Math.*, pages 1–13. Amer. Math. Soc., Providence, RI, 2003.

[3] E.B. Davies and B. Simon. Scattering theory for systems with different spatial asymptotics on the left and right. *Comm. Math. Phys.*, 63(3):277–301, 1978.

[4] M.S.P. Eastham. *The spectral theory of periodic differential equations.* Scottish Academic Press, Edinburgh/London, 1973.

[5] A. Figotin and P. Kuchment. Band-gap structure of spectra of periodic dielectric and acoustic media. I. Scalar model. *SIAM J. Appl. Math.*, 56(1):68–88, 1996.

[6] M. Sh. Birman and T.A. Suslina. A periodic magnetic Hamiltonian with a variable metric. The problem of absolute continuity. (Russian) *Algebra i Analiz* 11 (1999), no. 2, 1–40; translation in St. Petersburg Math. J. 11 (2000), no. 2, 203–232.

[7] R. Hempel and K. Lienau. Spectral properties of periodic media in the large coupling limit. *Comm. Partial Differential Equations*, 25(7-8):1445–1470, 2000.

[8] R. Hempel and O. Post. Spectral gaps for periodic elliptic operators with high contrast: an overview. In *Progress in analysis, Vol. I, II* (*Berlin,* 2001), pages 577–587. World Sci. Publ., River Edge, NJ, 2003.

[9] T. Kato: *Perturbation Theory for Linear Operators.* Springer, 1966.

[10] P. Kuchment. *Floquet theory for partial differential equations*, volume 60 of *Operator Theory: Advances and Applications.* Birkhäuser Verlag, Basel, 1993.

[11] P. Kuchment. The mathematics of photonic crystals. In *Mathematical modeling in optical science*, volume 22 of *Frontiers Appl. Math.*, pages 207–272. SIAM, Philadelphia, PA, 2001.

[12] M. Marino and A. Maugeri. L^p theory and partial Hölder continuity for quasilinear parabolic systems of higher order with strictly controlled growth. *Ann. di Math. Pura ed Appl.* **139**, 1 (1985) 107–145.

[13] F. Odeh and J.B. Keller. Partial differential equations with periodic coefficients and Bloch waves in crystals. *J. Mathematical Phys.*, 5:1499–1504, 1964.

[14] M. Reed and B. Simon. *Methods of modern mathematical physics. IV. Analysis of operators.* Academic Press [Harcourt Brace Jovanovich Publishers], New York, 1978.

[15] A. Tip, A. Moroz, and J.M. Combes. Band structure of absorptive photonic crystals. *J. Phys. A*, 33(35):6223–6252, 2000.

B.M. Brown
Cardiff School of Computer Science and Informatics
Cardiff University, Queen's Buildings, 5 The Parade
Roath, Cardiff CF24 3AA, UK
e-mail: `Malcolm.Brown@cs.cardiff.ac.uk`

V. Hoang and M. Plum
Karlsruhe Institute of Technology
Department of Mathematics
Institute for Analysis
D-76128 Karlsruhe, Germany
e-mail: `duy.hoang@kit.edu`
`michael.plum@kit.edu`

School of Mathematics, Statistics and Actuarial Science
University of Kent
Canterbury, Kent, CT2 7NZ, UK
e-mail: `i.wood@kent.ac.uk`

Operator Theory:
Advances and Applications, Vol. 214, 21–37
© 2011 Springer Basel AG

Representations of Compact Linear Operators in Banach Spaces and Nonlinear Eigenvalue Problems II

D.E. Edmunds, W.D. Evans and D.J. Harris

Abstract. This is a survey of recent work concerning a representation of compact linear operators acting between reflexive Banach spaces with strictly convex duals which is an analogue of Erhard Schmidt's classical Hilbert space theorem for compact operators.

Mathematics Subject Classification (2000). Primary 47A75; Secondary 47B06, 35P30.

Keywords. Compact linear operators, reflexive Banach spaces, strictly convex duals, duality maps, eigenvalues, p-Laplacian.

1. Introduction

A well-known classical result is that a compact self-adjoint operator A acting from a Hilbert space H into itself has the representation

$$Ax = \sum_n \lambda_n (x, \phi_n)\phi_n, \tag{1}$$

where the λ_n are eigenvalues of A, each repeated according to multiplicity and ordered by decreasing modulus, while the ϕ_n are orthonormal eigenvectors of A corresponding to the eigenvalues λ_n; the inner product in H is denoted by $(\cdot, \cdot)$. There is a finite number of eigenvalues if and only if A is of finite rank, otherwise the eigenvalues can accumulate only at zero. If the kernel of A is trivial then the eigenvectors ϕ_n form a complete orthonormal set in H. A consequence of this result is that if B is a compact linear operator mapping a Hilbert space H_1 into another Hilbert space H_2, then B has the representation

$$Bx = \sum_n \mu_n (x, \phi_n)_1 \psi_n, \tag{2}$$

where $(\cdot,\cdot)_1$ denotes the inner product in H_1, the ϕ_n are orthonormal eigenvectors of the positive square root $|B|$ of B^*B corresponding to eigenvalues μ_n (the μ_n are the *singular values* of B) and $\psi_n = \mu_n^{-1}B\phi_n$ ($\mu_n \neq 0$). All this may be found in [9], Chapter II, Section 5, for instance. Attempts have been made to generalise this result to cover the case of a compact linear operator T acting between Banach spaces X and Y. For instance, well-bounded operators were introduced by Smart [24] and an analogue of the result for compact self-adjoint operators on a Hilbert space established, namely, that a compact, well-bounded operator from a Banach space into itself can be represented by means of a sum of multiples of disjoint bounded projections: see [7] for details. This survey is based mainly on [11] where the objective was to deal with general compact linear operators, but to restrict the Banach spaces X and Y. It was shown that progress can be made if X and Y are assumed to be reflexive, strictly convex and have strictly convex duals. However, interesting new interpretations of the convergence results in [11] are given in Remarks 6 and 8, and Theorem 7 is new. In Theorem 9, quoted from [12], the analogue of (2) is presented for any compact linear map $T : X \to Y$, for reflexive Banach spaces X, Y with strictly convex duals.

The main results on the representations will be given in section 2, see Theorems 2 and 3: only sketch proofs are provided as details are available in [11]. These are intimately associated with a family of equations, indexed by a natural number k and involving the restrictions T_k of T to decreasing subspaces X_k of X, which are essentially the Euler equations obtained by maximising $\|T_kx\|_Y$ subject to the condition $\|x\|_{X_k} = 1$. Each of these equations has a solution x_k (an "eigenvector") corresponding to an "eigenvalue" λ_k. This has interesting applications. For example, it follows almost immediately that for each $k \in \mathbb{N}$, the Dirichlet eigenvalue problem for the p-Laplacian in a bounded open subset of $\mathbb{R}^n$ has a certain type of weak solution, called a k-weak solution, each such solution corresponding to an "eigenvalue" α_k, with $\alpha_k \to \infty$ as $k \to \infty$. A 1-weak solution is simply a classical weak solution, but for $k > 1$ the k-weak solutions are, in principle, more general, that is, weaker. Despite this, it is remarkable that the growth of α_k with k is exactly the same as that of the family of eigenvalues corresponding to classical weak solutions, the existence of which is commonly established by use of the heavier machinery of the Ljusternik-Schnirelmann category theory (see, for example, [8], [15] and also [3]).

Applications to eigenvalue problems for the Lane-Emden equation, the homogeneous p-Laplacian and Hardy-type operators are discussed in Section 3.

2. The representation theorems

We shall suppose throughout that X and Y are real, reflexive, strictly convex Banach spaces with strictly convex duals X^*, Y^* and norms $\|\cdot\|_X, \|\cdot\|_Y$: they are assumed to be real purely for ease of presentation. The value of $x^* \in X^*$ at $x \in X$ will be denoted by $\langle x, x^*\rangle_X$, and given any closed linear subspaces M, N of

X, X^*, respectively, we denote their *polar sets* by M^0, ${}^0N$; thus

$$M^0 = \{x^* \in X^* : \langle x, x^* \rangle_X = 0 \text{ for all } x \in M\},$$

and

$${}^0N = \{x \in X : \langle x, x^* \rangle_X = 0 \text{ for all } x^* \in N\}.$$

The following well-known notions and facts will feature prominently. They may be found in [6], [20] and [22].

- The polar set M^0 of a closed linear subspace of X is isometrically isomorphic to $(X/M)^*$;
- X is *strictly convex* if whenever $x, y \in X$ are such that $x \neq y$, $\|x\|_X = \|y\|_X = 1$ and $\lambda \in (0,1)$, then $\|\lambda x + (1-\lambda)y\|_X < 1$; equivalently, no sphere in X contains a line segment. X is *uniformly convex* if, for all $\varepsilon \in (0,2]$, $\|x\|_X = \|y\|_X = 1$ and $\|x - y\|_X \geq \varepsilon$, we have $\|x + y\|_X \leq 2(1 - \delta(\varepsilon))$, where the *modulus of convexity* of X

  $$\delta(\varepsilon) = \inf\{1 - \|x+y\|_X/2 : x, y \in X; \|x\|_X, \|y\|_X \leq 1, \|x-y\|_X \geq \varepsilon\} > 0.$$
- The norm $\|\cdot\|_X$ is Gâteaux differentiable on $X \setminus \{0\}$ if and only if X^* is strictly convex;
- The Gâteaux derivative $\widetilde{J}(x) := \text{grad}\|x\|_X$ of $\|x\|_X$ at $x \in X \setminus \{0\}$ is the unique element of X^* such that

  $$\left\|\widetilde{J}_X(x)\right\|_{X^*} = 1 \text{ and } \left\langle x, \widetilde{J}_X(x)\right\rangle_X = \|x\|_X;$$
- The map $J_X : X \to X^*$ defined by

  $$J_X(x) = \mu(\|x\|_X)\,\widetilde{J}_X(x) \quad (x \in X\backslash\{0\}),\ J_X(0) = 0, \tag{3}$$

  where $\mu : [0,\infty) \to [0,\infty)$ is a continuous, strictly increasing function with $\mu(0) = 0$ and $\lim_{t\to\infty} \mu(t) = \infty$, is called a *duality* map on X with gauge function μ. It satisfies

  $$\langle x, J_X x\rangle_X = \|J_X x\|_{X^*} \|x\|_X,\ \|J_X x\|_{X^*} = \mu(\|x\|_X). \tag{4}$$

  $J_X : X \to X^*$ is injective, surjective, strictly monotone and weakly continuous; the inverse of J_X is a duality map of X^* onto X with gauge function μ^{-1}. If X is uniformly convex, J_X is strongly continuous. We shall assume that the gauge function on X is normalised by $\mu(1) = 1$.
- When X is a Hilbert space, $\widetilde{J}_X(x) = x/\|x\|_X$, and we identify J_X (with gauge function $\mu(t) = t$) with the identity map of X to itself.

  The starting point is the following known result:

Proposition 1. *Let $T : X \to Y$ be compact and linear. Then there exists $x_1 \in X$, with $\|x_1\|_X = 1$, such that $\|T\| = \|Tx_1\|_Y$. Moreover, $x = x_1$ satisfies*

$$T^* \widetilde{J}_Y T x = \nu \widetilde{J}_X x, \tag{5}$$

with $\nu = \|T\|$; in terms of duality maps this equation has the form

$$T^* J_Y T x = \nu_1 J_X x,\ \nu_1 = \|T\|\,\mu_Y(\|T\|). \tag{6}$$

If $x \in X\backslash\{0\}$ satisfies (5) *for some ν, then $0 \le \nu \le \|T\|$ and $\|Tx\|_Y = \nu \|x\|_X$.*

Proof. The existence of x_1 is well known, and the fact that it satisfies (5) is merely a consequence of

$$\|T\| = \|Tx_1\|_Y = \max_{x \in X\backslash\{0\}} \frac{\|Tx\|_Y}{\|x\|_X},$$

since it then follows that for all $x \in X$,

$$\frac{d}{dt}\left(\frac{\|Tx_1 + tTx\|_Y}{\|x_1 + tx\|_X}\right)\bigg|_{t=0} = 0,$$

so that in terms of duality pairings,

$$\left\langle Tx, \widetilde{J}_Y Tx_1 \right\rangle_Y = \|Tx_1\|_Y \left\langle x, \widetilde{J}_X x_1 \right\rangle_X,$$

and hence

$$T^* \widetilde{J}_Y T x_1 = \lambda \widetilde{J}_X x_1,$$

with $\lambda = \|T\|$: cf. [2], and see also [5] and [17] for related results.

For the converse, let $x \in X\backslash\{0\}$ satisfy (5) for some ν. Then

$$\|Tx\|_Y = \left\langle Tx, \widetilde{J}_Y Tx \right\rangle_Y = \left\langle x, T^* \widetilde{J}_Y Tx \right\rangle_X = \nu \left\langle x, \widetilde{J}_X x \right\rangle_X = \nu \|x\|_X.$$

Hence $0 \le \nu \le \|T\|$. □

The equation (5) can be thought of as the Euler equation for maximising $\|Tx\|_Y$ subject to the condition $\|x\|_X = 1$, and $\|T\|$ is its maximum eigenvalue.

At this stage in the spectral analysis of T, when it is a compact self-adjoint operator acting on a Hilbert space H, one applies the result just established to the restriction of T to the orthogonal complement H_2 of x_1 in H, and then repeats the procedure to obtain successive orthogonal complements $H_k, k = 3, \ldots,$ the sequence of subspaces being infinite unless the restriction of T to one of them is the zero operator, in which case T is of finite rank. The absence of inner products when $T : X \to Y$ and X and Y are Banach spaces is a major hurdle, but the following procedure was shown to work in [11].

First set $X_1 = X, M_1 = \text{sp}\{J_X x_1\}$ (where sp denotes the linear span), $X_2 = {}^0M_1, N_1 = \text{sp}\{J_Y Tx_1\}, Y_2 = {}^0N_1$ and $\lambda_1 = \|T\|$. Since X_2 and Y_2 are closed subspaces of reflexive spaces they are reflexive. Also, $X_2^* = ({}^0M_1)^*$ is isometrically isomorphic to X_1^*/M_1; see [4], Proposition 9 in Section 5 of Chapter 1. From this fact and the assumed strict convexity of X^* it is proved in [11], Proposition 3, that X_2^* is strictly convex: the same argument applies to Y_2^*. Moreover, since by Proposition 1,

$$\langle Tx, J_Y Tx_1 \rangle_Y = \nu_1 \langle x, J_X x_1 \rangle_X \text{ for all } x \in X,$$

it follows that T maps X_2 to Y_2. The restriction T_2 of T to X_2 is thus a compact linear map from X_2 to Y_2, and if it is not the zero operator, by Proposition 1 there exists $x_2 \in X_2 \backslash \{0\}$ such that, with obvious notation,

$$\langle T_2 x, J_{Y_2} T_2 x_2 \rangle_{Y_2} = \nu_2 \langle x, J_{X_2} x_2 \rangle_{X_2} \text{ for all } x \in X_2,$$

where $\nu_2 = \lambda_2 \mu_Y(\lambda_2), \lambda_2 = \|Tx_2\|_Y = \|T_2\|$. Evidently $\lambda_2 \leq \lambda_1$ and $\nu_2 \leq \nu_1$. Continuing in this way we obtain elements $x_1, x_2, \ldots, x_n$ of X, all with unit norm, subspaces $M_1, \ldots, M_n$ of X^* and $N_1, \ldots, N_n$ of Y^*, where

$$M_k = \text{ sp } \{J_X x_1, \ldots, J_X x_k\} \text{ and } N_k = \text{ sp } \{J_Y T x_1, \ldots, J_Y T x_k\}, \; k = 1, \ldots, n,$$

and decreasing families $X_1, \ldots, X_n$ and $Y_1, \ldots, Y_n$ of subspaces of X and Y respectively given by

$$X_k = {}^0 M_{k-1}, Y_k = {}^0 N_{k-1}, \; k = 2, \ldots, n. \tag{7}$$

Moreover, for each $k \in \{1, \ldots, n\}$, T maps X_k into Y_k, $x_k \in X_k$ and with $T_k := T \upharpoonright_{X_k}, \lambda_k = \|T_k\| = \|Tx_k\|_Y, \nu_k = \lambda_k \mu_Y(\lambda_k)$, we have

$$\langle T_k x, J_{Y_k} T_k x_k \rangle_{Y_k} = \nu_k \langle x, J_{X_k} x_k \rangle_{X_k} \text{ for all } x \in X_k, \tag{8}$$

and so

$$T_k^* J_{Y_k} T_k x_k = \nu_k J_{X_k} x_k. \tag{9}$$

Note that (8) is equivalent to

$$\langle T_k x, J_Y T x_k \rangle_Y = \nu_k \langle x, J_X x_k \rangle_X, \quad x \in X_k. \tag{10}$$

For on identifying Y_k^* with the quotient space Y^*/Y_k^0, it follows that $J_{Y_k} y - J_Y y \in Y_k^0$ for any $y \in Y_k$ and hence, if $x \in X_k$,

$$\langle T_k x, J_{Y_k} y \rangle_{Y_k} = \langle T_k x, J_Y y \rangle_Y$$

since $T_k x \in Y_k$. Similarly for the right-hand sides of (8) and (10). Since $Tx_k \in Y_k = {}^0 N_{k-1}$, we have

$$\langle T x_k, J_Y T x_l \rangle_Y = 0 \text{ if } l < k. \tag{11}$$

The process stops with λ_n, x_n and X_{n+1} if and only if the restriction of T to X_{n+1} is the zero operator. In that case, the range of T is the linear space spanned by $Tx_1, \ldots, Tx_n$. For if $x \in X$, put

$$w_k = x - \sum_{j=1}^{k-1} \xi_j x_j, \quad \xi_j = \xi_j(x),$$

for $k \geq 2$, where the ξ_j are so chosen that $w_k \in X_k$. This choice is possible, and in a unique way, in view of (11): just take $\xi_1 = \langle x, J_X x_1 \rangle_X$ and for $2 \leq l \leq k-1$,

$$\xi_l = \left\langle x - \sum_{j=1}^{l-1} \xi_j x_j, J_X x_l, \right\rangle_X.$$

This means that $Tw_{n+1} = 0$, so that

$$Tx = \sum_{j=1}^{n} \xi_j T x_j = \sum_{j=1}^{n} \lambda_j \xi_j y_j, \text{ where } y_j = Tx_j / \|Tx_j\|_Y. \tag{12}$$

If T is not of finite rank then the sequence $\{\lambda_n\}$ is infinite and converges to zero. For then, since $Tx_n \in {}^0N_{n-1}$,

$$\left\langle Tx_n, \widetilde{J}_Y Tx_m \right\rangle_Y = 0 \text{ if } m < n. \tag{13}$$

Thus if $m < n$,

$$\begin{aligned} \lim_{k\to\infty} \lambda_k \le \|Tx_m\|_Y &= \left\langle Tx_m, \widetilde{J}_Y Tx_m \right\rangle_Y = \left\langle Tx_m - Tx_n, \widetilde{J}_Y Tx_m \right\rangle_Y \\ &\le \|Tx_m - Tx_n\|_Y \left\| \widetilde{J}_Y Tx_m \right\|_{Y^*} = \|Tx_m - Tx_n\|_Y . \end{aligned}$$

Since $\{x_n\}$ is bounded and T is compact, some subsequence of $\{Tx_n\}$ must converge and hence the assertion follows. Furthermore, if $x \in \cap_{n\in\mathbb{N}} X_n$, then, for all $n \in \mathbb{N}$, $\|Tx\|_Y \le \lambda_n \|x\|_X \to 0$ as $n \to \infty$, so that $\cap_{n\in\mathbb{N}} X_n$ is a subspace of the kernel $\ker(T)$ of T.

A key role is played by the family of maps

$$S_k : X \to \mathcal{M}'_{k-1} := \text{sp}\{x_1, \dots, x_{k-1}\}, k \ge 2,$$

determined by the condition that $x - S_k x \in X_k$ for all $x \in X$. By induction it follows that S_k is uniquely given by

$$S_k x := \sum_{j=1}^{k-1} \xi_j(x) x_j, \tag{14}$$

where, as noted above,

$$\xi_j(x) = \left\langle x - \sum_{i=1}^{j-1} \xi_i(x) x_i, J_X x_j \right\rangle_X \quad \text{for } j \ge 2, \quad \text{and} \quad \xi_1(x) = \langle x, J_X x_1 \rangle_X.$$

Hence S_k is linear. Also from the uniqueness, it follows that $S_k^2 = S_k$ and S_k is a linear projection of X onto $\mathcal{M}'_{k-1}$. In fact, it is established in [11], Proposition 9 that for each $k \ge 2, X$ and X^* have the direct sum decompositions

$$X = X_k \oplus \mathcal{M}'_{k-1}, \quad X^* = M_{k-1} \oplus (\mathcal{M}'_{k-1})^0. \tag{15}$$

The operators S_k, S_k^* are respectively linear projections of X onto $\mathcal{M}'_{k-1}$ and X^* onto M_{k-1}. The identity (10) can therefore be written as

$$\langle T(I - S_k)x, J_Y Tx_k \rangle_Y = \nu_k \langle (I - S_k)x, J_X x_k \rangle_X, \quad \text{for all } x \in X. \tag{16}$$

Hence

$$(T^* J_Y T - \nu_k J_X)\, x_k \in X_k^0 = M_{k-1},$$

and

$$(I^* - S_k^*)\,(T^* J_Y T - \nu_k J_X)\, x_k = 0. \tag{17}$$

Given any $x \in X$ and $k \in \mathbb{N}$, let z_k be the point in X_k nearest to x, so that $z_k = P_k x$, where $P_k = P_{X_k}$ is the (in general, non-linear) projection of X onto X_k. Then as $\|z_k - x\|_X \le \|x\|_X$ it follows that

$$\|z_k\|_X \le 2\, \|x\|_X \tag{18}$$

and of course

$$\|Tz_k\|_Y \leq \lambda_k \|z_k\|_X . \tag{19}$$

Since

$$\|z_k - x\|_X = \inf\{\|x - z_k + ty\|_X : y \in X_k\},$$

we see that for all $y \in X_k$,

$$\frac{d}{dt}\|x - z_k + ty\|_X \Big|_{t=0} = 0,$$

and hence

$$\langle y, J_X(x - z_k)\rangle_X = 0 \text{ for all } y \in X_k. \tag{20}$$

It follows that $J_X(x - z_k) \in M_{k-1}$, and thus, by (15), we have

$$J_X(x - P_k x) = S_k^* J_X(x - P_k x)$$

and so

$$x - P_k x = J_X^{-1} S_k^* J_X(x - P_k x). \tag{21}$$

Furthermore, since $\lambda_k \to 0$ as $k \to \infty$, we have on using (18) and (19),

$$Tx = \lim_{k\to\infty} T J_X^{-1} S_k^* J_X(x - P_k x). \tag{22}$$

If T has infinite rank, there is a strictly increasing sequence $(k(j))_{j\in\mathbb{N}}$ of natural numbers such that the weak limit

$$w - \lim_{j\to\infty} z_{k(j)}$$

exists and it lies in $\ker T$ since $\cap_{k\in\mathbb{N}} X_k \subseteq \ker T$. If $\ker T = \{0\}$, then $z_{k(j)} \to 0$ weakly and

$$x = w - \lim_{j\to\infty} J_X^{-1} S_{k(j)}^* J_X(x - P_{k(j)} x). \tag{23}$$

This can be expressed in the form

$$x = w - \lim_{j\to\infty} J_X^{-1} \left(\sum_{i=1}^{k(j)-1} \eta_i(k(j), x) J_X x_i \right),$$

for some real constants $\eta_i(k(j), x)$. However, a deeper analysis yields the following results in [11], Theorem 17, Remarks 18 and 19, Theorem 21 and Corollary 22:

Theorem 2. *Suppose that X is uniformly convex and X^* is strictly convex; put $X_\infty = \cap_{k\in\mathbb{N}} X_k$ and write P_k, P_∞ for the projections onto X_k, X_∞, respectively. Then for all $x \in X$, $P_k x \to P_\infty x$ as $k \to \infty$ and*

$$x = \lim_{k\to\infty} (I - P_k) S_k x + P_\infty x. \tag{24}$$

This can be expressed as

$$x = \lim_{k\to\infty} \left\{ J_X^{-1} \sum_{j=1}^{k-1} \eta_j(k, x) J_X x_j \right\} + P_\infty x,$$

for some real constants $\eta_j(k,x)$. *If* $\ker(T) = \{0\}$ *and* $\lim_{k\to\infty} S_k x$ *exists, then* $x = \sum_{j=1}^{\infty} \xi_j(x) x_j$.

If X *is a Hilbert space, then for any* $x \in X$,

$$x = \sum_{j=1}^{\infty} (x, x_j)_X x_j + P_\infty x.$$

From (24) it follows that

$$Tx = \lim_{k\to\infty} T(I - P_k) S_k x \tag{25}$$

since $X_\infty \subseteq \ker T$. If X is a Hilbert space then

$$Tx = \sum_{j=1}^{\infty} (x, x_j)_X T x_j.$$

In [11], Remark 20, it was noted that the operator T was only needed in Theorem 2 to establish the existence of the sequence $\{x_j\}$ with the appropriate properties. This has the following implication for a representation of Tx in terms of the sequence $\{y_j\}$ in Y, where $y_j = Tx_j/\|Tx_j\|_X = \lambda_j^{-1} T x_j$. The analogues of the maps S_k are maps $R_k : Y \to \mathcal{N}'_{k-1} := \mathrm{sp}\{y_1, y_2, \ldots, y_{k-1}\}, k \geq 2$, determined by the conditions $y - R_k y \in Y_k$ for all $y \in Y$, and we have the direct sum decomposition

$$Y = \mathcal{N}'_{k-1} \oplus Y_k.$$

The R_k are linear projections onto $\mathcal{N}'_{k-1}$ and are uniquely given by

$$R_k y = \sum_{j=1}^{k-1} \gamma_j(y) y_j,$$

where

$$\gamma_j(y) = \left\langle y - \sum_{i=1}^{j-1} \gamma_i(y) y_i, J_Y y_j \right\rangle_Y, \quad \gamma_1(y) = \langle y, J_Y y_1 \rangle_Y.$$

It follows that $\gamma_j(Tx) = \lambda_j \xi_j(x)$ and $R_k(Tx) = TS_k x$. The resulting theorem is

Theorem 3. *Suppose that* Y *is uniformly convex and* Y^* *is strictly convex; put* $Y_\infty = \cap_{k\in\mathbb{N}} Y_k$ *and write* Q_k, Q_∞ *for the (generally nonlinear) projections of* Y *onto* Y_k, Y_∞ *respectively. Then for all* $x \in X, Q_k T x \to Q_\infty T x$ *as* $k \to \infty$ *and*

$$Tx = \lim_{k\to\infty} (I - Q_k) T S_k x + Q_\infty T x, \tag{26}$$

where $TS_k x = \sum_{j=1}^{k-1} \xi_j(x) T x_j = \sum_{j=1}^{k-1} \lambda_j \xi_j(x) y_j$.

If Y *is a Hilbert space, then for all* $x \in X$,

$$Tx = \sum_{j=1}^{\infty} \lambda_j \xi_j y_j + Q_\infty T x, \tag{27}$$

where $\xi_j = \lambda_j^{-1} (Tx, y_j)_Y$.

Remark 4. When X and Y are Hilbert spaces, the duality maps are identified with the identity and the duals with the original spaces; also $S_k^* = S_k$. The direct sums in (15) are now orthogonal sums and $T^*Tx_k \in X_k$. Hence $S_k(T^*T - \nu_k I)x_k = 0$. On using (17), this yields $|T|^2 x_k := T^*Tx_k = \lambda_k^2 x_k$, since $\nu_k = \lambda_k^2$. Hence λ_k is a singular value of T. In this case, Theorem 3 gives the classical Schmidt representation of T and T^*, and $Y_\infty = \{0\}$, (see, for example, Chapter II, §5 of [9]).

It is shown in [12], Theorem 6, that the final term in (27) is redundant (and hence $Y_\infty = \{0\}$) for all Banach spaces X and Y with strictly convex duals, (see Theorem 9 below).

Remark 5. Let X be a Hilbert space and $\ker(T) = \{0\}$. We have shown in Theorem 2 that, for all $x \in X$,

$$x = \sum_j (x, x_j)_X x_j.$$

Hence, since $T^*J_YTx \in X$, we have

$$\begin{aligned} T^*J_YTx &= \sum_j (T^*J_YTx, x_j)_X x_j \\ &= \sum_j \langle Tx_j, J_YTx\rangle_Y x_j. \end{aligned}$$

In particular, since $Y_j = {}^0N_{j-1} \subseteq {}^0N_l$ if $j-1 \geq l$,

$$\begin{aligned} T^*J_YTx_l &= \sum_{j\leq l} \langle Tx_j, J_YTx_l\rangle_Y x_j \\ &= \langle Tx_l, J_YTx_l\rangle_Y x_l + \sum_{j\leq l-1} \langle Tx_j, J_YTx_l\rangle_Y x_j \\ &= \nu_l\langle x_l, J_Xx_l\rangle_X x_l + \sum_{j\leq l-1} \langle Tx_j, J_YTx_l\rangle_Y x_j \end{aligned}$$

by (10). Since $\langle x_l, J_Xx_l\rangle_X = \|x_l\|_X = 1$ it follows that

$$T^*J_YTx_l = \nu_l x_l + \sum_{j\leq l-1} \langle Tx_j, J_YTx_l\rangle_Y x_j. \tag{28}$$

Thus, as we already know, $T^*J_YTx_l = \nu_l x_l$ if $l = 1$ or Y is a Hilbert space, but is the converse true?

Remark 6. It is illuminating to interpret Theorem 2 in the following way. First define

$$\begin{aligned} S := \Bigg\{ z \in \prod_{k\in\mathbb{N}} (X/X_k) : z = (z_k),\, z_k = \phi_k(z_{k+1}) \\ \text{and } \|z\|_S := \sup_k \left(\|z_k\|_{X/X_k}\right) < \infty \Bigg\}, \end{aligned} \tag{29}$$

where X/X_k are quotient spaces and $\phi_k : X/X_{k+1} \to X/X_k$ is the canonical map. To simplify notation, we omit the canonical maps of X into X/X_∞ and X into X/X_k, whenever the meaning is unambiguous.

From (15), $\mathcal{M}'_{k-1} \simeq X/X_k$, and the map $S_k : X \to \mathcal{M}'_{k-1}$ in (14) satisfies

$$\|S_k x\|_{X/X_k} = \|S_k x - P_k S_k x\|_X. \tag{30}$$

With $u = S_k x$, we have from (15) that $x - u \in X_k$ and so $P_k(u - x) = u - x$. Also, since $P_k x$ is the unique element $w \in X_k$ for which $\|x - w\|_X$ is minimal, we infer from

$$\|x - (x - u + P_k u)\|_X = \|u - P_k u\|_X$$

that $P_k x = x - u + P_k u$, which gives

$$P_k S_k x - P_k x = S_k x - x$$

and hence

$$P_k S_k x - S_k x = P_k x - x.$$

On substituting in (30), we have

$$\|S_k x\|_{X/X_k} = \|x - P_k x\|_X = \|x\|_{X/X_k}. \tag{31}$$

Since $\|\cdot\|_{X/X_k}$ increases with k, it follows that

$$\sup_k \|S_k x\|_{X/X_k} = \lim_{k\to\infty} \|S_k x\|_{X/X_k} = \|x\|_{X/X_\infty}, \tag{32}$$

where $X_\infty := \cap_{k=1}^{\infty} X_k$, this having been shown to be a subspace of the kernel of T.

For $x \in X$, define

$$\Phi : x \mapsto (S_k x) = \left(\sum_{i=1}^{k-1} \xi_i(x) x_i \right).$$

On continuing our abuse of notation by writing x_i for $\phi_k(x_i)$ if $i \le k$, we can assert from (32) that Φ maps X into S and is an isometry if we assume that $X_\infty = \{0\}$. We now show that Φ maps X onto S.

Let $(v_k) \in S$, that is, v_k is in the conjugacy class

$$\sum_{i=1}^{k-1} \xi_i x_i + X_k$$

and

$$\sup_k \left\| \sum_{i=1}^{k-1} \xi_i x_i \right\|_{X/X_k} < \infty. \tag{33}$$

Choose v_k to be the element of the conjugacy class of minimum norm, that is

$$\|v_k\|_X = \left\| \sum_{i=1}^{k-1} \xi_i x_i \right\|_{X/X_k}.$$

Then, by (33), $\{v_k\}$ is bounded in X and so contains a weakly convergent subsequence (which we continue to denote by $\{v_k\}$) with weak limit v, say. For $k > j, \phi_j(v_k)$ is constant, namely $\sum_{i=1}^{j-1} \xi_i x_i$, since the spaces X_k are decreasing. Moreover, the canonical maps are weakly continuous. Hence $\phi_j(v_k) \to \phi_j(v)$ as $k \to \infty$, and $\phi_j(v) = \sum_{i=1}^{j-1} \xi_i x_i \in \mathcal{M}'_{j-1}$. Thus $(v_j) = \Phi v$ as claimed.

We have therefore proved

Theorem 7. *Let S denote the space of formal series $\sum_{i=1}^{\infty} \xi_i x_i$ with norm*

$$\sup_k \left\| \sum_{i=1}^{k-1} \xi_i x_i \right\|_{X/X_k} < \infty.$$

and suppose $X_\infty = \{0\}$. Then the map $\Phi : x \mapsto \left(\sum_{i=1}^{k-1} \xi_i(x) x_i\right) : X \to S$ is an isometric isomorphism. If X is a separable Hilbert space, the result is just the Parseval theorem.

If $X_\infty \neq \{0\}$, we can work with X/X_∞ instead of X. Theorem 7 then holds with Φ an isometric isomorphism of X/X_∞ onto S.

Similarly, with W the space of formal series $\sum_1^\infty \gamma_i y_i$ with norm

$$\sup_k \left\| \sum_{i=1}^{k-1} \gamma_i y_i \right\|_{Y/Y_k},$$

the map $\Psi : y \mapsto \left(\sum_{i=1}^{k-1} \gamma_i(y) y_i\right) : Y/Y_\infty \to W$ is an isometric isomorphism and

$$Tx \sim \left(\sum_{i=1}^{k-1} \gamma_i(Tx) y_i \right)$$

Remark 8. Suppose that $X_\infty = \{0\}$. Then

$$S_n : X \to \mathcal{M}'_{n-1} = \mathrm{sp}\{x_1, x_2, \ldots, x_{n-1}\} \simeq X/X_n$$

implies that

$$\bigcap_{n \in \mathbb{N}} S_n^{-1}(\{0\}) = \bigcap_{n \in \mathbb{N}} X_n = \{0\}.$$

The *projective limit* topology on X is the coarsest topology compatible with the algebraic structure of X under which all the maps S_n are continuous, and this is a locally convex topology; see [23]. If $\mathcal{V}_n := \{v_n\}$ is a base of absolutely convex neighbourhoods in $\mathcal{M}'_{n-1}$, then finite intersections of $S_n^{-1} v_n$ $(v_n \in \mathcal{V}_n)$ form a base of absolutely convex neighbourhoods of X in the projective limit topology.

We have $v_n = B_\varepsilon(0) \cap \mathcal{M}'_{n-1}$ for some $\varepsilon > 0$, where $B_\varepsilon(0)$ is the ball centre 0 and radius ε defined with respect to the norm induced by X (which is equivalent to any other norm since $\mathcal{M}'_{n-1}$ is of finite dimension). Since

$$S_n^{-1}\left(B_\varepsilon(0) \cap \mathcal{M}'_{n-1}\right) = \left\{B_\varepsilon(0) \cap \mathcal{M}'_{n-1} + X_n\right\}$$

and

$$\begin{aligned}
&\left\{B_\varepsilon(0) \cap \mathcal{M}'_{n-1} + X_n\right\} \bigcap \left\{B_\delta(0) \cap \mathcal{M}'_{k-1} + X_k\right\} \\
&\quad \supseteq \left\{B_{\min\{\varepsilon,\delta\}}(0) \cap \mathcal{M}'_{\min\{n,k\}-1} + X_{\max\{n,k\}}\right\},
\end{aligned}$$

a base of neighbourhoods of X is given by sets of the form

$$\left\{B_\varepsilon(0) \cap \mathcal{M}'_{m-1} + X_k : \varepsilon > 0, k \geq m\right\}.$$

Given $\varepsilon > 0, k \in \mathbb{N}$, then, for $n \geq k$,

$$S_n(x) - x \in X_n \subseteq B_\varepsilon(0) \cap \mathcal{M}'_{m-1} + X_k.$$

Therefore $S_n x \to x$ in the projective limit topology on X.

The argument at the end of Remark 6 leads to a proof of the following analogue of the Schmidt representation (2) established in [12]. However, we refer the reader to the direct proof given in [12], Theorem 6, which is more transparent and instructive.

Theorem 9. *Let X, Y be reflexive Banach spaces with strictly convex duals and $T : X \to Y$ a compact linear operator. Then, for all $x \in X, TS_n x \to Tx$ in Y, that is,*

$$Tx = \sum_{i=1}^{\infty} \lambda_i \xi_i(x) y_i, \quad y_i := Tx_i / \|Tx_i\|_Y.$$

If T is of finite rank, the sum is finite.

To conclude this section we consider the case in which $X = Y$, so that $T : X \to X$ is a compact linear map with spectrum that, apart from the point 0, consists solely of classical eigenvalues of finite algebraic multiplicity. Let $\{\widetilde{\lambda}_n\}$ be the sequence of all non-zero classical eigenvalues of T, repeated according to algebraic multiplicity and ordered so that

$$\left|\widetilde{\lambda}_1\right| \geq \left|\widetilde{\lambda}_2\right| \geq \cdots \geq 0.$$

If T has only m $(< \infty)$ distinct classical eigenvalues and M is the sum of their algebraic multiplicities, we put $\widetilde{\lambda}_n = 0$ for all $n > M$. Our concern here is the relationships between the 'eigenvalues' λ_n discussed earlier in this section, the classical eigenvalues $\widetilde{\lambda}_n$ and various s-numbers: we recall that s-numbers form sequences of real numbers used to help determine the 'degree of compactness' of maps. These numbers are defined for any bounded linear map $S : E \to F$, where E and F are Banach spaces; the ones we shall need are:

(i) the approximation numbers $a_n(S)$, where

$$a_n(S) := \inf \|S - L\| \quad (n \in \mathbb{N}),$$

the infimum being taken over all bounded linear maps $L : E \to F$ with rank $L := \dim L(E) < n$;

(ii) the Gelfand numbers $c_n(S)$, where

$$c_n(S) := \inf \|S \mid_{\widetilde{E}}\|,$$

and the infimum is taken over all linear subspaces $\widetilde{E}$ of E with codim $\widetilde{E} < n$;

(iii) the Weyl numbers $x_n(S)$, where

$$x_n(S) := \sup\{a_n(SA) : \|A : l_2 \to E\| \leq 1\}.$$

Details of the main properties of these numbers may be found in [9], [18] and [21].

First we observe that if $\widetilde{\lambda}$ is a non-zero eigenvalue of T corresponding to a normalised eigenvector x, then

$$\left\langle x, \lambda^{-1}T^*\widetilde{J_X x}\right\rangle_X = \left\langle \lambda^{-1}Tx, \widetilde{J_X x}\right\rangle_X = \left\langle x, \widetilde{J_X x}\right\rangle_X = 1,$$

so that by the strict convexity and reflexivity of X^*, we have

$$T^*\widetilde{J_X x} = \lambda \widetilde{J_X x}.$$

This means that $\widetilde{J_X x}$ is an eigenvector of T^* with corresponding eigenvalue λ; by (3), so is $J_X x$. Moreover, since $\widetilde{J_X}(\lambda x) = (\text{sgn } \lambda)\widetilde{J_X x}$,

$$T^*\widetilde{J_X}Tx = T^*\widetilde{J_X}(\lambda x) = (\text{sgn } \lambda)T^*\widetilde{J_X x} = |\lambda|\, \widetilde{J_X x},$$

so that the eigenvector x of T satisfies our basic equation (5). Consideration of suitable compact Volterra integral operators shows that solutions of (5) need not be eigenvectors of T.

Next, since

$$\text{codim } X_k = \text{ dim } (\text{sp}\{J_X x_1, \dots, J_X x_{k-1}\}),$$

it follows immediately that

$$c_n(T) \le \lambda_n \ (n \in \mathbb{N}).$$

Moreover,

$$x_n(T) \le c_n(T) \le a_n(T) \ (n \in \mathbb{N})$$

(see [21], p. 115), and

$$\left|\widetilde{\lambda}_{2n-1}\right| \le e \left(\prod_{k=1}^{n} x_k(T) \right)^{1/n}$$

(see [21], p. 156). Hence

$$\left|\widetilde{\lambda}_{2n-1}\right| \le e \left(\prod_{k=1}^{n} c_k(T) \right)^{1/n} \le e \left\|T\right\|^{(n-1)/n} c_n^{1/n}(T) \le e \left\|T\right\|^{(n-1)/n} \lambda_n^{1/n}.$$

However, while this gives some connection between our eigenvalues and the classical ones, it does not seem particularly useful or sharp.

3. Applications

Let Ω be a bounded open subset of $\mathbb{R}^n$, let $1 < p < \infty$ and let $W_p^1(\Omega)$ be the Sobolev space of all real-valued functions $u \in L_p(\Omega)$ all of whose first-order distributional derivatives $D_j u$ also belong to $L_p(\Omega)$. The norm on $W_p^1(\Omega)$ is defined to be

$$\left(\int_\Omega \left\{ |u|^p + \sum_{j=1}^{n} |D_j u|^p \right\} dx \right)^{1/p}.$$

We take X to be $\overset{0}{W}{}_p^1(\Omega)$, the closure in $W_p^1(\Omega)$ of the set $C_0^\infty(\Omega)$ of all infinitely differentiable functions with compact support in Ω, and define the norm on X by

$$\|u\|_X = \left(\int_\Omega \sum_{j=1}^n |D_j u|^p \, dx \right)^{1/p}. \tag{34}$$

Because of the Friedrichs inequality (see [9], Theorem V.3.22), this norm is equivalent to the norm on X inherited from $W_p^1(\Omega)$. Let $Y = L_p(\Omega)$, $T = \mathrm{id} : X \to Y$; id is compact. It is plain that both X and Y are reflexive and strictly convex. The strict convexity of Y^* is obvious, while that of X^*, follows easily from the Gâteaux-differentiability of $\|\cdot\|_X$ on $X\backslash\{0\}$. Direct verification shows that

$$\widetilde{J}_Y u = \|u\|_p^{-(p-1)} |u|^{p-2} u, \tag{35}$$

where $\|\cdot\|_p$ is the usual norm on $L_p(\Omega)$. As for $\widetilde{J}_X$, we claim that

$$\widetilde{J}_X u = - \|u\|_X^{-(p-1)} \Delta_p u \text{ in the sense of distributions,} \tag{36}$$

where

$$\Delta_p u = \sum_{j=1}^n D_j \left(|D_j u|^{p-2} D_j u \right), \tag{37}$$

corresponding to a version of the p-Laplacian. To verify this, note that for all $u \in X$,

$$\begin{aligned} \left\langle u, - \|u\|_X^{-(p-1)} \Delta_p u \right\rangle_X &= - \|u\|_X^{-(p-1)} \langle u, \Delta_p u \rangle_X \\ &= \|u\|_X^{-(p-1)} \int_\Omega \sum_{j=1}^n D_j u. |D_j u|^{p-2} D_j u dx \\ &= \|u\|_X . \end{aligned}$$

With $\mu_X(t) = \mu_Y(t) = t^{p-1}$, the corresponding duality maps J_X, J_Y are given by

$$J_X(u) = -\Delta_p u, \quad J_Y(u) = |u|^{p-2} u.$$

Our basic eigenvalue equation (5) then gives

$$\left\langle v, \widetilde{J}_Y u_1 \right\rangle_Y = \lambda_1 \left\langle v, \widetilde{J}_X u_1 \right\rangle_X \text{ for all } v \in X,$$

which, since $\|u_1\|_Y = \|\mathrm{id}\| = \lambda_1$, amounts to

$$\int_\Omega v |u_1|^{p-2} u_1 dx = \lambda_1^p \int_\Omega \sum_{j=1}^n (D_j v) |D_j u_1|^{p-2} D_j u_1 dx, \tag{38}$$

so that u_1 is a weak solution of the Dirichlet eigenvalue problem

$$-\Delta_p u_1 = \lambda_1^{-p} |u_1|^{p-2} u_1, \; u_1 = 0 \text{ on } \partial\Omega. \tag{39}$$

Since id is not of finite rank, our general procedure ensures that for each $k \in \mathbb{N}$, there are an "eigenvector" u_k and a corresponding "eigenvalue" λ_k^{-p} that satisfy

$$-\Delta_p u_k = \lambda_k^{-p}|u_k|^{p-2}u_k, \quad u_k = 0 \text{ on } \partial\Omega, \tag{40}$$

in the sense that for all $v \in X_k$,

$$\int_\Omega v|u_k|^{p-2}u_k d\mathbf{x} = \lambda_k^p \int_\Omega \sum_{j=1}^n (D_j v)|D_j u_k|^{p-2} D_j u_k d\mathbf{x}. \tag{41}$$

We shall refer to u_k as a k-weak solution of (40): note that when $k = 1$, all functions in $X_1 = X$ are allowed as test functions, while for general $k > 1$ the test functions have to be taken from $X_k \subset X$. The smoothness of these test functions remains unclear to us. Results of this kind are known for the eigenvalues obtained by the Lyusternik-Schnirel'mann procedure (see, for example, [8] and [15]), but as remarked in the Introduction, the simplicity of the present approach has its attractions. Moreover, information about the growth of the λ_k^{-p} can be obtained quite painlessly. In fact, if we let id : $W_p^1(\Omega) \to L_p(\Omega)$ be the natural embedding, then from the definitions of the kth Gelfand and Weyl numbers we have

$$\lambda_k \geq c_k(\mathrm{id}_k) \geq c_k(\mathrm{id}) \geq x_k(\mathrm{id}).$$

From [18], Theorem 3.c.5 and Remark 3.c.7 (1), we see that $x_k(\mathrm{id}) \geq ck^{-1/n}$, where c is a positive constant independent of k. Hence $\lambda_k \geq ck^{-1/n}$, and so the eigenvalues λ_k^{-p} of (40) are $0(k^{p/n})$. This upper estimate of the growth of the eigenvalues is exactly that obtained for the Lyusternik-Schnirel'mann eigenvalues in [14] and [16] (these corresponding to classical weak solutions), where lower bounds of the same order are also established.

We recall that the mth such eigenvalue, which we denote by $\widehat{\lambda}_m$, is given by (see [1], Theorem 3.4 and [25], Chapter 44)

$$\widehat{\lambda}_m = \inf_{K \in \mathbb{A}_m} \sup_{u \in K} \|u\|_p^{-p},$$

where $\mathbb{A}_m$ is the family of all compact, symmetric subsets K of $\{u \in \overset{0}{W}{}_p^1(\Omega) : \|\nabla u\|_p = 1\}$ with genus $\gamma(K) \geq m$, the genus being defined as

$$\gamma(K) = \inf\{k \in \mathbb{N} : \text{there is a continuous, odd map } h : K \to \mathbb{R}^k \backslash \{0\}\}.$$

The corresponding quantities λ_m^{-p} obtained by our method are expressible as

$$\lambda_m^{-p} = \inf_{u \in X_m \backslash \{0\}} \frac{\|\nabla u\|_p^p}{\|u\|_p^p}.$$

It is not clear what connection there may be between the eigenvalues found by our method and those given by the Lyusternik-Schnirel'mann procedure, nor whether there are eigenvalues not found by either process.

If instead we suppose that $1 < p < n$, $q \in \left(1, \frac{np}{n-p}\right)$ and take $X = \overset{0}{W}{}_p^1(\Omega)$, $Y = L_q(\Omega)$, $T = \mathrm{id} : X \to Y$, with Ω as before, then as id is compact, the same procedure establishes the existence of a countable family of eigenvectors v_k and eigenvalues μ_k of the p, q-Laplacian Dirichlet eigenvalue problem

$$-\Delta_p v = \mu |v|^{q-2} v, \; v = 0 \text{ on } \partial\Omega, \tag{42}$$

where each v_k is a k-weak solution of (42) in the sense just described. The same holds when $n = 1$, $1 < p < \infty$ and $1 < q < \infty$ since, by [19], Remarks 5.8.4 (i), $\overset{0}{W}{}^1_p(\Omega)$ is compactly embedded in $C(\overline{\Omega})$ and hence in $L_q(\Omega)$. Note that the special case $p = 2$ of (42) is of physical interest as it involves the Lane-Emden equation, of importance in astrophysics.

Applications to more complicated differential operators, including ones of higher-order, can be found in [11], Section 4. Another natural application is to Hardy-type operators of the form

$$(Tf)(x) = v(x) \int_a^x u(t) f(t) dt, \; x \in I = [a, b] \subset \mathbb{R}$$

considered as a map from $L_p(I)$ into $L_q(I)$, where $p, q \in (1, \infty)$, and u, v are given positive functions in $L_{p'}(I), L_q(I)$ respectively, with $p' = p/(p-1)$. Under these conditions it is known that T is compact (see, for example, [10], Chapter 2). Our results give immediately that for each $k \in \mathbb{N}$, there is an "eigenvalue" λ_k of the basic equation (5), with $\lambda_k \to \infty$ as $k \to \infty$. In the case $p = q$, it is shown in [2], that the corresponding Euler equation (5) has an infinite set of eigenvalues and that these eigenvalues coincide with the approximation numbers of T. Furthermore, an asymptotic formula for the eigenvalues is derived; see [13] for the general case.

References

[1] An Lê, Eigenvalue problems for the p-Laplacian, Nonlinear Analysis **64** (2006), 1057–1099.

[2] Bennewitz, C., Approximation numbers = singular eigenvalues, J. Comp. and Appl. Math., **208** (2007), 102–110.

[3] Binding, P., Boulton, L., Čepička, J., Drábek, P. and Girg, P., Basis properties of the p-Laplacian, Proc. American Math. Soc. **134** (2006), 3487–3494.

[4] Bourbaki, N., *Espaces vectoriels topologiques*, Hermann, Paris, 1956.

[5] Boyd, D.W., Inequalities for positive integral operators, Pacific J. of Math. **38** (1971), 9–24.

[6] Browder, F.E., Nonlinear operators and nonlinear equations of evolution in Banach spaces, Proc. AMS Symp. Pure Mathematics, Vol. XVIII, Part 2, AMS, Providence, Rhode Island, 1976.

[7] Cheng Qingping and Doust, I., Compact well-bounded operators, Glasgow Math. J. **43** (2001), 467–475.

[8] Coffman, C.V., Lyusternik-Schnirelman theory and eigenvalue problems for monotone potential operators, J. Funct. Anal. **14** (1973), 237–252.

[9] Edmunds, D.E. and Evans, W.D., *Spectral theory and differential operators*, Oxford University Press, Oxford, 1987.

[10] Edmunds, D.E. and Evans, W.D., *Hardy operators, function spaces and embeddings*, Springer, Berlin-Heidelberg-New York, 2004.

[11] Edmunds, D.E., Evans, W.D. and Harris, D.J., Representations of compact linear operators in Banach spaces and nonlinear eigenvalue problems, J. London Math. Soc. **78** (2008), 65–84.

[12] Edmunds, D.E., Evans, W.D. and Harris, D.J., A spectral analysis of compact linear operators in Banach spaces, B. London Math. Soc. **42** (4) (2010), 726–734.

[13] Edmunds, D.E. and Lang, J., Asymptotics for eigenvalues of a non-linear integral system, Boll. U.M.I. **1** (Series IX) (2008), 105–120.

[14] Friedlander, L., Asymptotic behavior of the eigenvalues of the p-Laplacian, Commun. Partial Diff. Equations **14** (1989), 1059–1069.

[15] Garcia Azorero, J.P. and Peral Alonso, I., Existence and non-uniqueness for the p-Laplacian: nonlinear eigenvalues, Comm. Partial Differential Equations **12** (1987), 1389–1430.

[16] Garcia Azorero, J.P. and Peral Alonso, I., Comportement asymptotique des valeurs propres du p-laplacien, C.R. Acad. Sci. Paris **301** (1988), 75–78.

[17] Howard, R. and Shep, A.R., Norms of positive operators on L_p spaces, Proc. American Math. Soc. **109** (1990), 135–146.

[18] König, H., *Eigenvalue distribution of compact operators*, Birkhäuser, Basel, 1986.

[19] Kufner, A., John, O. and Fučik, S., *Function spaces,* Academia, Prague, 1977.

[20] Lions, J.-L., *Quelques méthodes de résolution des problèmes aux limites non linéaires,* Dunod, Paris, 1969.

[21] Pietsch, A., *Eigenvalues and s-numbers*, Cambridge University Press, Cambridge, 1987.

[22] Martin, R.H., *Nonlinear operators and differential equations in Banach spaces*, Wiley, New York, 1976.

[23] Robertson, A.P. and Robertson, W.J., *Topological Vector Spaces,* Cambridge University Press, Cambridge, 1964.

[24] Smart, D.R., Conditionally convergent spectral expansions, J. Austral. Math. Soc. Ser. A **1** (1960), 319–333.

[25] Zeidler, *Nonlinear functional analysis and its applications,* Springer, Berlin, 1990.

D.E. Edmunds
Department of Mathematics, Sussex University
Brighton, BN1 9RF, England, UK
e-mail: `DavidEEdmunds@aol.com`

W.D. Evans and D.J. Harris
School of Mathematics, Cardiff University
Senghennydd Road
Cardiff, CF24 4AG, Wales, UK
e-mail: `EvansWD@cardiff.ac.uk`
`pryske@boyns.net`

Operator Theory:
Advances and Applications, Vol. 214, 39–44
© 2011 Springer Basel AG

A Sharp Bound on Eigenvalues of Schrödinger Operators on the Half-line with Complex-valued Potentials

Rupert L. Frank, Ari Laptev and Robert Seiringer

Abstract. We derive a sharp bound on the location of non-positive eigenvalues of Schrödinger operators on the half-line with complex-valued potentials.

Mathematics Subject Classification (2000). 35P15, 81Q12.

Keywords. Schrödinger operator, complex potential, eigenvalue bounds.

1. Introduction and main result

In this note we are concerned with estimates for non-positive eigenvalues of one-dimensional Schrödinger operators with complex-valued potentials. We shall provide an example of a bound where the sharp constant *worsens* when a Dirichlet boundary condition is imposed. This is in contrast to the case of real-valued potentials, where the variational principle implies that the absolute value of the non-positive eigenvalues decreases.

In order to describe our result, we first assume that V is real-valued. It is a well-known fact (attributed to L. Spruch in [K]) that any negative eigenvalue λ of the Schrödinger operator $-\partial^2 - V$ in $L^2(\mathbb{R})$ satisfies

$$|\lambda|^{1/2} \leq \frac{1}{2} \int_{-\infty}^{\infty} |V(x)| \, \mathrm{d}x \,. \tag{1.1}$$

The constant $\frac{1}{2}$ in this inequality is sharp and attained if $V(x) = c\delta(x-b)$ for any $c > 0$ and $b \in \mathbb{R}$. (It follows from the Sobolev embedding theorem that the operator $-\partial^2 - V$ can be defined in the quadratic form sense as long as V is a finite Borel measure on $\mathbb{R}$. In this case the right side of (1.1) denotes the total variation of the measure.) From (1.1) and the variational principle for self-adjoint operators

Support through DFG grant FR 2664/1-1 (R.F.) and U.S. National Science Foundation grants PHY 0652854 (R.F.) and PHY 0652356 (R.S.) is gratefully acknowledged.

we immediately infer that any negative eigenvalue of the operator $-\partial^2 - V$ in $L^2(0,\infty)$ with Dirichlet boundary conditions satisfies

$$|\lambda|^{1/2} \leq \frac{1}{2} \int_0^\infty |V(x)|\, dx\,. \tag{1.2}$$

The constant $\frac{1}{2}$ in this inequality is still sharp but no longer attained.

Motivated by concrete physical examples and problems in computational mathematics, an increasing interest in eigenvalue estimates for *complex-valued* potentials has developed in recent years. A beautiful observation of [AAD] is that (1.1) remains valid for all eigenvalues in $\mathbb{C} \setminus [0,\infty)$ even if V is complex-valued. The same is not true for (1.2) ! Indeed, our main result is

Theorem 1.1. *For $a \in \mathbb{R}$ let*

$$g(a) := \sup_{y \geq 0} \left| e^{iay} - e^{-y} \right| . \tag{1.3}$$

Any eigenvalue $\lambda = |\lambda| e^{i\theta} \in \mathbb{C} \setminus [0,\infty)$ of the operator $-\partial^2 - V$ in $L^2(0,\infty)$ with Dirichlet boundary conditions satisfies

$$|\lambda|^{1/2} \leq \frac{1}{2}\, g(\cot(\theta/2)) \int_0^\infty |V(x)|\, dx\,. \tag{1.4}$$

This bound is sharp in the following sense: For any given $m > 0$ and $\theta \in (0, 2\pi)$ there are $c \in \mathbb{C}$ and $b > 0$ such that for $V(x) = c\delta(x-b)$ one has $|c| = \int |V(x)|\, dx = m$ and the unique eigenvalue of $-\partial^2 - V$ is given by $(m^2/4)\, g(\cot(\theta/2))^2 e^{i\theta}$, that is, equality is attained in (1.4).

Remark 1.2. Our bound does not apply to positive eigenvalues. In the case of real-valued potential it is known that there are no positive eigenvalues if $V \in L^1(\mathbb{R})$.

We note that $1 < g(a) < 2$ for $a > 0$. The following lemma discusses the function g in more detail.

Lemma 1.3. *For $a \geq 0$, the function $g(a)$ is monotone increasing, with $g(0) = 1$ and $\lim_{a\to\infty} g(a) = 2$. Moreover,*

$$g(a) = 1 + O(e^{-\pi/(3a)}) \tag{1.5}$$

for small a, and

$$g(a) = 2 - \frac{\pi}{a} + O(a^{-2}) \tag{1.6}$$

as $a \to \infty$.

In Figure 1 we plot the curve $\{|z| = g(\cot(\theta/2))^2\}$. It follows from (1.6) that this curve hits the positive real axis at the point 4 with slope $2/\pi$. Close to the point -1 the curve coincides with a semi-circle up to exponentially small terms, as (1.5) shows.

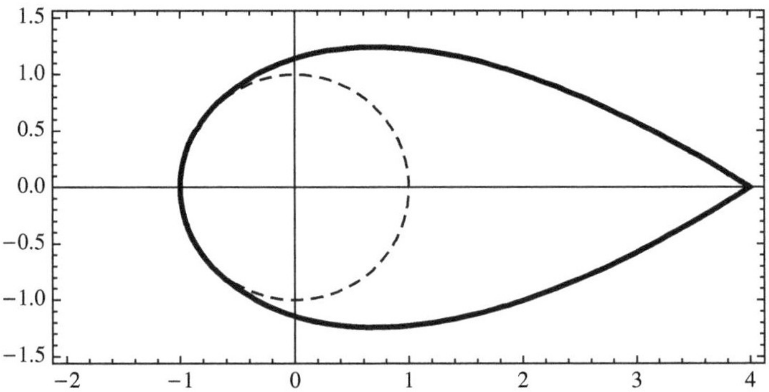

FIGURE 1. The maximal value of $4|\lambda|$ on the half-line with $\int_0^\infty |V(x)|\,\mathrm{d}x = 1$. The dashed line is the corresponding bound on the whole line.

Using that $\sup_a g(a) = 2$ we find

Corollary 1.4. *Any eigenvalue $\lambda \in \mathbb{C} \setminus [0,\infty)$ of the operator $-\partial^2 - V$ in $L^2(0,\infty)$ with Dirichlet boundary conditions satisfies*

$$|\lambda|^{1/2} \le \int_0^\infty |V(x)|\,\mathrm{d}x\,. \tag{1.7}$$

The bound is not true in general if the right side is multiplied by a constant < 1.

Inequality (1.7) follows also from inequality (1.1) for complex-valued potentials. Indeed, the odd extension of an eigenfunction of the Dirichlet operator is an eigenfunction of the whole-line operator with the potential $V(|x|)$ with the same eigenvalue. The remarkable fact is that the inequality is sharp in the complex-valued case, as shown in Theorem 1.1.

By the same argument (1.7) is also valid if *Neumann* instead of Dirichlet boundary conditions are imposed. In this case equality holds for any $V(x) = c\delta(x)$ with $\operatorname{Re} c > 0$. In particular, in the Neumann case (1.7) is sharp for any fixed argument $0 < \theta < 2\pi$ of the eigenvalue λ. The analogue for mixed boundary conditions is

Proposition 1.5. *Let $\sigma \ge 0$. Any eigenvalue $\lambda \in \mathbb{C} \setminus [0,\infty)$ of the operator $-\partial^2 - V$ in $L^2(0,\infty)$ with boundary conditions $\psi'(0) = \sigma\psi(0)$ satisfies*

$$|\lambda|^{1/2} \le \int_0^\infty |V(x)|\,\mathrm{d}x\,. \tag{1.8}$$

The bound is sharp for any $\sigma \ge 0$ and any fixed argument $0 < \theta < 2\pi$ of the eigenvalue λ.

Note that if $\sigma < 0$ a bound of the form (1.8) can not hold since there exists a non-positive eigenvalue even in the case $V = 0$.

Remark 1.6. In the *self-adjoint* case inequality (1.1) for whole-line operators is accompanied by bounds

$$|\lambda|^\gamma \leq \frac{\Gamma(\gamma+1)}{\sqrt{\pi}\,\Gamma(\gamma+3/2)}\left(\frac{\gamma-1/2}{\gamma+1/2}\right)^{\gamma-1/2}\int_{-\infty}^{\infty}|V(x)|^{\gamma+1/2}\,\mathrm{d}x \tag{1.9}$$

for $\gamma > 1/2$; see [K, LT]. In contrast, in the *non-selfadjoint* case it seems to be unknown whether the condition $V \in L^{\gamma+1/2}(\mathbb{R})$ for some $1/2 < \gamma < \infty$ implies that all eigenvalues in $\mathbb{C}\setminus[0,\infty)$ lie inside a finite disc; see [DN, FLLS, LS, S] for partial results in this direction. We would like to remark here that even if a bound of the form (1.9) were true in the non-selfadjoint case with $1/2 < \gamma < \infty$, then (in contrast to (1.1) for $\gamma = 1/2$) the constant would have to be strictly larger than in the self-adjoint case. To see this, consider $V(x) = \frac{\alpha(\alpha+1)}{\cosh^2 x}$ with $\operatorname{Re}\alpha > 0$. Then $\lambda = -\alpha^2$ is an eigenvalue (with eigenfunction $(\cosh x)^{-\alpha}$) and the supremum

$$\sup_{\operatorname{Re}\alpha\geq 0}\frac{|\lambda|^\gamma}{\int_{-\infty}^{\infty}|V(x)|^{\gamma+1/2}\,\mathrm{d}x} = \left(\int_{-\infty}^{\infty}\frac{dx}{\cosh^2 x}\right)^{-1}\sup_{\operatorname{Re}\alpha\geq 0}\frac{|\alpha|^{\gamma-1/2}}{|\alpha+1|^{\gamma+1/2}}$$

is clearly attained for purely imaginary values of α.

2. Proofs

Proof of Theorem 1.1. Assume that $-\partial^2\psi(x) - V(x)\psi(x) = -\mu\psi(x)$ with $\psi(0) = 0$, $\psi \not\equiv 0$ and $\mu = -\lambda \in \mathbb{C}\setminus(-\infty,0]$. Then the Birman-Schwinger operator

$$V^{1/2}\frac{1}{-\partial^2+\mu}|V|^{1/2}, \qquad V^{1/2} := (\operatorname{sgn} V)|V|^{1/2},$$

has an eigenvalue 1, and hence its operator norm is greater or equal to 1.

The integral kernel of this operator equals

$$V(x)^{1/2}\frac{\mathrm{e}^{-\sqrt{\mu}|x-y|} - \mathrm{e}^{-\sqrt{\mu}(x+y)}}{2\sqrt{\mu}}|V(y)|^{1/2},$$

and hence

$$\left|\left(\psi, V^{1/2}\frac{1}{-\partial^2+\mu}|V|^{1/2}\varphi\right)\right| \leq \frac{\|V\|_1}{2\sqrt{|\mu|}}\|\psi\|_2\|\varphi\|_2\sup_{x,y\geq 0}\left|\mathrm{e}^{-\sqrt{\mu}|x-y|} - \mathrm{e}^{-\sqrt{\mu}(x+y)}\right|.$$

Without loss of generality, we can take the supremum over the smaller set $x \geq y \geq 0$. Then

$$\sup_{x\geq y\geq 0}\left|\mathrm{e}^{-\sqrt{\mu}(x-y)} - \mathrm{e}^{-\sqrt{\mu}(x+y)}\right| = \sup_{x\geq y\geq 0}\mathrm{e}^{-x\operatorname{Re}\sqrt{\mu}}\left|\mathrm{e}^{\sqrt{\mu}y} - \mathrm{e}^{-\sqrt{\mu}y}\right|.$$

Since $\operatorname{Re}\sqrt{\mu} > 0$, the supremum over x is achieved at $x = y$, and hence

$$\sup_{x,y\geq 0}\left|\mathrm{e}^{-\sqrt{\mu}(x-y)} - \mathrm{e}^{-\sqrt{\mu}(x+y)}\right| = \sup_{y\geq 0}\left|1 - \mathrm{e}^{-2\sqrt{\mu}y}\right|.$$

If we write $\mu = -|\mu|\,\mathrm{e}^{i\theta}$ with $0 < \theta < 2\pi$, then

$$\sup_{y\geq 0}\left|1 - \mathrm{e}^{-2\sqrt{\mu}y}\right| = \sup_{y\geq 0}\left|\mathrm{e}^{2i\sqrt{|\mu|}\cos(\theta/2)y} - \mathrm{e}^{-2\sqrt{|\mu|}\sin(\theta/2)y}\right| = g(\cot(\theta/2))$$

with g from (1.3). Hence we have shown that

$$\left\|V^{1/2}\frac{1}{-\partial^2+\mu}|V|^{1/2}\right\| \leq \frac{\|V\|_1}{2\sqrt{|\mu|}}\, g(\cot(\theta/2))\,. \tag{2.1}$$

Since the left side is greater or equal to 1, as remarked above, we obtain (1.4).

For $V(x) = c\delta(x-b)$ the Birman-Schwinger operator reduces to the number $c(1-e^{-2\sqrt{\mu}b})/(2\sqrt{\mu})$ and inequality (2.1) becomes equality provided $\sqrt{\mu}b$ satisfies $|1-e^{-2\sqrt{\mu}b}| = g(\cot(\theta/2))$. For given $m > 0$ and $\theta \in (0, 2\pi)$ this determines b and $|c|$. The phase of c is found from the equation $c(1-e^{-2\sqrt{\mu}b})/(2\sqrt{\mu}) = 1$. □

Proof of Lemma 1.3. By continuity for $a > 0$ there exists an optimizer y_0 such that $g(a) = |\,\mathrm{e}^{iay_0} - \mathrm{e}^{-y_0}\,|$. We claim that y_0 satisfies $\pi/3 < ay_0 \leq \pi$. To see the lower bound, note that $|\,\mathrm{e}^{iay} - \mathrm{e}^{-y}\,| \geq 1$ if and only if $2\cos(ay) \leq \mathrm{e}^{-y}$. In particular, $\cos(ay_0) < 1/2$. For the upper bound, if $2\pi > ay > \pi$ and $2\cos(ay) < \mathrm{e}^{-y}$, replacing ya by $2\pi - ya$ leads to a contradiction. Similarly, if $ya > 2\pi$ it can replaced by $ya - 2\pi$ in order exclude that y is the optimizer.

It is elementary to check that $|\,\mathrm{e}^{iay} - \mathrm{e}^{-y}\,|$ is monotone increasing in a for every fixed y with $0 \leq y \leq \pi/a$. Since we know already that $y_0 \leq \pi/a$, the monotonicity of g follows.

Plugging in $y = \pi/a$, we obtain $g(a) \geq 1 + \mathrm{e}^{-\pi/a} \geq 2 - \pi/a$. For large enough a, it follows from this that y_0 is close to π/a. In particular, $y_0 \geq \pi/(2a)$, and hence $|\,\mathrm{e}^{iay_0} - 1| \geq g(a) \geq 2 - \pi/a$. This implies that $y_0 = \pi/a + O(a^{-2})$, and thus $g(a) = 2 - \pi/a + O(a^{-2})$, as claimed.

For an upper bound for small a, we use the triangle inequality and the bound $ay_0 \geq \pi/3$ to find $g(a) \leq 1 + e^{-y_0} \leq 1 + \mathrm{e}^{-\pi/(3a)}$. □

Proof of Proposition 1.5. We proceed as in the proof of Theorem 1.1. The Birman-Schwinger operator has the kernel

$$V(x)^{1/2}\frac{\mathrm{e}^{-\sqrt{\mu}|x-y|} + \frac{\sqrt{\mu}-\sigma}{\sqrt{\mu}+\sigma}\,\mathrm{e}^{-\sqrt{\mu}(x+y)}}{2\sqrt{\mu}}|V(y)|^{1/2}\,.$$

The assertion follows as above using that

$$\sup_{y\geq 0}\left|1 + \frac{\sqrt{\mu}-\sigma}{\sqrt{\mu}+\sigma}\,\mathrm{e}^{-2\sqrt{\mu}y}\right| \leq 2$$

by the triangle inequality and the fact that $|\sqrt{\mu}-\sigma| \leq |\sqrt{\mu}+\sigma|$. The fact that the bound (1.8) is sharp for given argument $0 < \theta < 2\pi$ of the eigenvalue λ follows by choosing $V(x) = -ci\,\mathrm{e}^{i\theta/2}\,\delta(x)$ for $c > 0$ and letting $c \to \infty$. □

References

[AAD] A.A. Abramov, A. Aslanyan, E.B. Davies, *Bounds on complex eigenvalues and resonances.* J. Phys. A **34** (2001), no. 1, 57–72.

[DN] E.B. Davies, J. Nath, *Schrödinger operators with slowly decaying potentials.* J. Comput. Appl. Math. **148** (2002), no. 1, 1–28.

[FLLS] R.L. Frank, A. Laptev, E.H. Lieb, R. Seiringer, *Lieb-Thirring inequalities for Schrödinger operators with complex-valued potentials.* Lett. Math. Phys. **77** (2006), no. 3, 309–316.

[K] J.B. Keller, *Lower bounds and isoperimetric inequalities for eigenvalues of the Schrödinger equation.* J. Mathematical Phys. **2** (1961), 262–266.

[LS] A. Laptev, O. Safronov, *Eigenvalue estimates for Schrödinger operators with complex potentials.* Commun. Math. Phys. **292** (2009), no. 1, 29–54.

[LT] E.H. Lieb, W. Thirring, *Inequalities for the moments of the eigenvalues of the Schrödinger Hamiltonian and their relation to Sobolev inequalities.* In: Studies in Mathematical Physics, E.H. Lieb, et al. (eds.), 269–303. Princeton Univ. Press, Princeton, NJ, 1976.

[S] O. Safronov, *Estimates for eigenvalues of the Schrödinger operator with a complex potential.* Preprint arXiv:0902.3950v1 (2009).

Rupert L. Frank
Department of Mathematics
Princeton University
Princeton, NJ 08544, USA
e-mail: `rlfrank@math.princeton.edu`

Ari Laptev
Department of Mathematics
Imperial College London
London SW7 2AZ, UK

and

Department of Mathematics
Royal Institute of Technology
S-100 44 Stockholm, Sweden
e-mail: `a.laptev@imperial.ac.uk`
`laptev@math.kth.se`

Robert Seiringer
Department of Physics
Princeton University
Princeton, NJ 08544, USA
e-mail: `rseiring@princeton.edu`

Operator Theory:
Advances and Applications, Vol. 214, 45–66
© 2011 Springer Basel AG

Zero-range Model of *p*-scattering by a Potential Well

Yu.E. Karpeshina

Abstract. A well-known method of zero-range potentials consists of replacing a deep potential well of a small radius by a boundary condition at the point of the centre of the well. However, in passing to the limit from a deep and narrow potential well to the zero-range model, information, concerning *p*-scattering and scatterings of higher orders, disappears. Traditional zero-range model describes only bound states and s-scattering. The principal mathematical difficulty, which arises in the mathematical construction of a zero-range model, describing *p*-scattering, is that *p*-scattered waves have a square non-integrable singularity at the point, where the well should be located. It is not possible to construct the corresponding energy operator in $L_2(\mathbb{R}^3)$. We construct the energy operator in some Hilbert space, which naturally arises from the problem and includes $L_2(\mathbb{R}^3)$. We explicitly construct the complete system of generalized eigenfunctions in this space.

Mathematics Subject Classification (2000). 35Q40, 81V45.

Keywords. Zero-range potential, *p*-scattering.

1. Introduction

Behavior of scattered waves at large distances from a potential well is often described by the following expansion:

$$\Psi(k,x) =_{|x|\to\infty} \Psi_0(k,x), \tag{1.1}$$

$$\Psi_0(k,x) = e^{i(k,x)} + \sum_{n\in N_0^3} \nu_n(k)\frac{\partial^{|n|}}{\partial x^n}\frac{e^{i|k||x|}}{|x|}, \tag{1.2}$$

where $\nu_n(k)$ are scattering amplitudes,

$$N_0 = \{0,1,2,\dots\}, \quad n=(n_1,n_2,n_3), \quad \frac{\partial^{|n|}}{\partial x^n} = \frac{\partial^{n_1+n_2+n_3}}{\partial x_1^{n_1}\partial x_2^{n_2}\partial x_3^{n_3}}.$$

The term corresponding to $n = (0,0,0)$ in (1.2) is called S-scattered wave, the terms corresponding to the first derivatives are P-scattered waves. A basic problem in Scattering Theory is to connect characteristics of the well with scattering amplitudes.

There is a method of zero-range potential, widely used in Quantum Mechanics, which connects properties of the well with the S-scattered wave. The method consists of replacing a deep potential well of a small radius by a boundary condition at the centre of the well, [1]–[3]. Namely, let

$$H_V = -\Delta + V, \tag{1.3}$$

where

$$V(x) = \begin{cases} -E_0, & |x| \leq r_0, \\ 0, & |x| > r_0. \end{cases} \tag{1.4}$$

Let r_0 tends to zero, $E_0 > 0$, E_0 is of order r_0^{-2}. It is possible to choose $E_0(r_0)$ in such a way that H has a fixed negative eigenvalue $-\alpha^2$ for all sufficiently small r_0. The corresponding eigenfunction is close to the Green function $ce^{-\alpha|x|}|x|^{-1}$ outside the well. This observation motivated physicists to replace H_V by an operator $H(\alpha)$, which acts in $L_2(\mathbb{R}^3)$ and is described by the formula:

$$H(\alpha)u(x)_{|x|\neq 0} = -\Delta u(x). \tag{1.5}$$

The domain of $H(\alpha)$ is the set of functions u which admit the representation $u = u_0 + ce^{-\alpha|x|}/|x|$, where $u_0 \in W_2^{2,0}$, $c \in C$, $W_2^{2,0}$ being the closure in W_2^2 of the set of infinitely differentiable functions with the support in $\mathbb{R}^3 \setminus \{0\}$. By embedding theorem [4], u_0 is continuous and $u_0(0) = 0$. Hence, $u(x)$ satisfies the following asymptotic near the origin:

$$u(x) =_{|x|\to 0} c(|x|^{-1} - \alpha) + o(1). \tag{1.6}$$

The first rigorous description of $H(\alpha)$ is given in [5]. The family of operators $H(\alpha)$, $\alpha \in \mathbb{R}$, is a family of self-adjoint extensions of a symmetric operator $H_0 = -\Delta$, $D(H_0) = W_2^{2,0}$ acting in $L_2(\mathbb{R}^3)$. The operator H_0 has the deficiency indices equal to $(1,1)$. Its self-adjoint extensions can be described by von Neumann formulas [6] and turned out to coincide with the family $H(\alpha)$, $\alpha \in \mathbb{R} \cup \{\infty\}$. Spectral study of $H(\alpha)$ is rather simple. The negative spectrum of $H(\alpha)$ consists of an eigenvalue $-\alpha^2$ if $\alpha > 0$ and empty if $\alpha \leq 0$. The spectrum of $H(\alpha)$ includes the positive semi-axis $[0, \infty)$. The positive part of the spectrum is absolutely continuous. Generalized eigenfunctions have a form of scattered plane waves:

$$\Psi(k,x) = e^{i(k,x)} + \nu_0(k)e^{i|k||x|}/|x|, \quad k \in \mathbb{R}^3,$$

where the coefficient $\nu_0(k)$ is defined by the asymptotic (1.6):

$$\nu_0 = -\frac{1}{\alpha + i|k|}.$$

Different configurations of such "atoms" are widely used in physics to describe atomic systems (see, e.g., [2, 3]).

A shortcoming of the zero-range model is that it describes only spherically symmetric scattered waves $\nu_0(k)e^{i|k||x|}/|x|$ (S-scattering). In passing to the limit from a deep and narrow potential well to the zero-range model, information, concerning p-scattering and scatterings of higher orders on the well, vanishes.

The principal mathematical difficulty, which arises in the rigorous construction of a zero-range potential, describing p-scattering, is that a p-scattered wave is a combination of the first derivatives of $e^{ik|x|}/|x|$. The derivatives have a singularity at the point $x = 0$, which is not square integrable. For this reason it is not possible to construct an energy operator in $L_2(\mathbb{R}^3)$.

There are a lot of different approaches to constructing an operator, describing P-scattering from a narrow well. Some of them rely on using standard Hilbert spaces. For example, the space of functions being square integrable with the functional weight $\mid x \mid^{\beta}$, $\beta > 1/2$ is considered in [7]. However, Laplacian is not symmetric in this space and has to be replaced by $\mid x \mid^{-\beta} \Delta$. Eigenfunctions of this operator don't obey the Helmholtz equation, which is expected to describe behavior of eigenfunctions in the whole space $\mathbb{R}^3$ except the region of the well. In the case of other standard Hilbert spaces similar problems arise. Energy operator turns out to be trivial (just the Laplacian) or not self-adjoint or its eigenfunctions don't satisfy Helmholtz equation. Other approaches include an indefinite metric approach, see, e.g., [8]–[20]. The main problem, arising in the indefinite metric approach is that the constructed Hamiltonian is not a self-adjoint operator in a Hilbert space. Its spectrum may have non-real eigenvalues. One more approach has been developed in [21], where a potential well is replaced by a semitransparent sphere with a condition on a boundary jump of the normal derivative. In this case variables turn out to be separable and, therefore, the problem can be explicitly solved. However, since this model is not zero-range, solutions are difficult to get in the case of several wells.

The model constructed here is essentially described long time ago in [22]. Since then new publications have appeared. It was discovered in [24]–[26] that it is possible to determine P-scattering through self-adjoint operators in a certain extended Hilbert space. The method of "triplet extensions" was developed in [27], [28].

We suggest an approach different from those developed until now. The operator constructed here has the following properties. First of all, it is zero-range: its eigenfunctions satisfy Helmholtz equation for all $x \in \mathbb{R}^3, x \neq 0$. Next, eigenfunctions have the form of p-scattered waves. Finally, the energy operator is self-adjoint in some Hilbert space, which naturally arises from the problem and includes $L_2(\mathbb{R}^3)$. p-scattered waves form a complete orthogonal system in this space.

The scheme of the construction is the following. First, we introduce a set of functions $\psi(k, x)$, $k \in \mathbb{R}^3$, which have a form of p-scattered waves, satisfying the Helmholtz equation in $\mathbb{R}^3 \setminus \{0\}$, and have the prescribed asymptotic behavior, when $x \to 0$, see (see (2.4)). A positive parameter α is included in this asymptotic formula. Different values of the parameter correspond to different wells, which are approximated by the zero-range model. Next, we describe explicitly the Hilbert space, where the system of these p-scattered waves is complete and orthogonal. After that, we define the Hamiltonian as the operation of multiplication by the variable $\mid k \mid^2$ in this system. Finally, we show that this Hamiltonian acts on functions from its domain simply as a Laplacian everywhere, except the point $x = 0$.

In Section 6 we present a boundary conditions and the operator corresponding to a spherically symmetric well. It should be noted that following this scheme one can consider scattering of higher orders and consider other elliptic systems, f.e. for Lamé equations. Furthermore, this model can be easily generalized to cases of many centers and, therefore, be used for the study of scattering by complicated systems of wells.

2. *P*-scattered waves and boundary conditions

Since a potential is "concentrated" at the origin, we expect that P- scattered waves to have a form:

$$\psi(k,x) = (2\pi)^{-3/2} \exp i(k,x) + \nu(k) G'_{|k|^2+i0}(x) \tag{2.1}$$

$$G'_{|k|^2+i0}(x) = \frac{\partial}{\partial x_1} \frac{\exp ik \mid x \mid}{\mid x \mid}, \tag{2.2}$$

with some coefficient $\nu(k) \in C$. It is easy to see that $\psi(k,x)$ has the following asymptotic expansion near the origin:

$$\psi(k,x) =_{x \to 0} \nu(k) \frac{\partial \mid x \mid^{-1}}{\partial x_1} + \tag{2.3}$$

$$+ b(k) \frac{\partial \mid x \mid}{\partial x_1} + (2\pi)^{-3/2} + a(k) x_1 + i(2\pi)^{-3/2} k_2 x_2 + i(2\pi)^{-3/2} k_3 x_3 + o(\mid x \mid),$$

$$a(k) = i(2\pi)^{-3/2} k_1 + \nu(k)(i|k|)^3/3, \quad b(k) = -\frac{k^2 \nu(k)}{2}.$$

By analogy with (1.6), we choose a condition connecting coefficients in front of the singularity and in front of x_1. Namely, $\beta \nu(k) = a(k)$, where β is a negative coefficient, the same for all $k \in \mathbb{R}^3$. It is convenient (we will see this later) to write β in the form $\beta = (-\alpha)^3/3$, $\alpha > 0$. Thus, $\psi(k,x)$ has an asymptotic:

$$\psi(k,x) =_{x \to 0} \nu(k) \left(\frac{\partial \mid x \mid^{-1}}{\partial x_1} - \frac{\alpha^3}{3} x_1 \right) + b(k) \frac{\partial \mid x \mid}{\partial x_1} \tag{2.4}$$

$$+ (2\pi)^{-3/2} + i(2\pi)^{-3/2} k_2 x_2 + i(2\pi)^{-3/2} k_3 x_3 + o(\mid x \mid).$$

We immediately get formula for $\nu(\vec{k})$ from the equation $a(k) = \alpha^3 \nu(k)/3$:

$$\nu(k) = -\frac{(2\pi)^{-3/2} 3ik_1}{\alpha^3 + (i \mid k \mid)^3}. \tag{2.5}$$

We do not have physical motivation for choosing condition (2.4). To our best knowledge, the boundary conditions at zero, corresponding to a small inclusion are not established; here by inclusion we mean a potential well or elastic inclusion.

The difficulty in establishing such a connection is that a direct computation should hold for all values of $k \in \mathbb{R}^3$. The purpose of the paper is to describe a new approach in constructing a self-adjoint operator, corresponding to boundary conditions (1) and waves $\psi(k, x)$.

3. Space $\mathcal{H}_\alpha$ and Operator A_α

To begin with we consider the Hilbert space $\mathcal{H}_\alpha$ of the functions, which can be represented in the form:

$$F = f + \gamma G'_{-\alpha^2}(x), \tag{3.1}$$

where $f \in L_2(\mathbb{R}^3), \gamma \in C$,

$$G'_{-\alpha^2}(x) = \frac{\partial}{\partial x_1} \frac{\exp(-\alpha \mid x \mid)}{\mid x \mid}, \tag{3.2}$$

α being a positive value fixed for a given $\mathcal{H}_\alpha$. A scalar product in $\mathcal{H}_\alpha$ is determined as follows:

$$[F_1, F_2] = (f_1, f_2) + 2\pi\alpha\gamma_1\overline{\gamma_2}, \tag{3.3}$$

$(\cdot, \cdot)$, being the scalar product in $L_2(\mathbb{R}^3)$. We show, that $\mathcal{H}_\alpha$ can be constructed as the closed span of $\psi(k, x)$, $k \in \mathbb{R}^3$.

Note that all the spaces $\mathcal{H}_\alpha$, $\alpha > 0$ may be defined by the same set of functions $\mathcal{H}$:

$$\mathcal{H} = \left\{ F(x) = f(x) + \gamma \left(\frac{\partial}{\partial x_1} \frac{1}{|x|} \right) \chi(x), \quad f \in L_2(\mathbb{R}^3), \quad \gamma \in \mathbf{C} \right\}$$

χ being the characteristic function of the unit ball centered at the origin. Spaces $\mathcal{H}_\alpha$ differ by their scalar product. Note that the difference is not trivial, the scalar products differ more essentially than just by a constant multiplier. In fact,

$$\|F\|_{H_{\alpha_1}} = \|F\|_{H_{\alpha_2}}, \text{ if } F \in L_2(\mathbb{R}^3), \tag{3.4}$$

for any pair α_1, α_2, $\| \cdot \|_{H_\alpha}$ being the norm in $\mathcal{H}_\alpha$. However,

$$\|G'_{-\alpha_1^2}\|_{H_{\alpha_1}} = \sqrt{2\pi\alpha_1},$$
$$\|G'_{-\alpha_2^2}\|_{H_{\alpha_2}} = \sqrt{2\pi\alpha_2}.$$

Note that $G'_{-\alpha_1^2} - G'_{-\alpha_2^2} \in L_2(\mathbb{R}^3)$, since it has a weaker singularity than $G'_{-\alpha_1^2}$ or $G'_{-\alpha_2^2}$. By (3.4),

$$\begin{aligned} \|G'_{-\alpha_1^2} - G'_{-\alpha_2^2}\|_{H_{\alpha_1}} &= \|G'_{-\alpha_1^2} - G'_{-\alpha_2^2}\|_{H_{\alpha_2}} \\ &= \|G'_{-\alpha_1^2} - G'_{-\alpha_2^2}\|_{L_2}^2 = \frac{(\alpha_1 - \alpha_2)^2}{12\pi^2(\alpha_1 + \alpha_2)}. \end{aligned}$$

Therefore for any F being a linear combination of $G'_{-\alpha_1^2}, G'_{-\alpha_2^2}$, $F = \nu_1 G'_{-\alpha_1^2} + \nu_2 G'_{-\alpha_2^2}$, we have the following norms:

$$\|F\|^2_{H_{\alpha_1}} = 2\pi\alpha_1(\nu_1+\nu_2)^2 + \nu_2^2 \frac{(\alpha_1-\alpha_2)^2}{12\pi^2(\alpha_1+\alpha_2)},$$
$$\|F\|^2_{H_{\alpha_2}} = 2\pi\alpha_2(\nu_1+\nu_2)^2 + \nu_1^2 \frac{(\alpha_1-\alpha_2)^2}{12\pi^2(\alpha_1+\alpha_2)}.$$

A function $\psi(k,x)$ has oscillatory behavior at infinity and therefore does not belong to $\mathcal{H}_\alpha$. However, $\psi(k,x)\varphi_0 \in \mathcal{H}_\alpha$ for any $\varphi_0 \in C_0^\infty(\mathbb{R}^3)$, here and below $C_0^\infty(\mathbb{R}^3)$ is the set of infinitely differentiable functions with compact support. It is easy to show that $\psi(k,x)\varphi_0(x) = f_\Psi(k,x) + \nu(k)G'_{-\alpha^2}(x)$, where $f_\Psi(k,x) \in L_2(\mathbb{R}^3)$,

$$\begin{aligned} f_\Psi(k,x) &= (2\pi)^{-3/2}\exp i(k,x)\varphi_0(x) \\ &\quad + \nu(k)\left(G'_{|k|^2+i0}(x)\varphi_0(x) - G'_{-\alpha^2}(x)\right). \end{aligned} \tag{3.5}$$

Let $\mathcal{H}^0_\alpha$ be the set of functions F in $\mathcal{H}_\alpha$ whose $L_2(\mathbb{R}^3)$-component $f(x)$ belongs to $C_0^\infty(\mathbb{R}^3)$. Obviously, $\mathcal{H}^0_\alpha$ is dense in $\mathcal{H}_\alpha$. We consider the scalar product $[F,\psi(k)]$, $F \in \mathcal{H}^0_\alpha$. Strictly speaking the scalar product is not yet defined, since $\psi(k,x) \notin \mathcal{H}_\alpha$. However, $[F,\psi(k)\varphi_0]$ is defined for any $\varphi_0 \in C_0^\infty(\mathbb{R}^3)$. Indeed, it is easy to show that

$$[F,\psi(k)\varphi_0] = \int_{\mathbb{R}^3} f(x) f_\Psi(k,x)dx + 2\pi\alpha\gamma\bar{\nu}$$

for any $\varphi_0(x)$ equal to 1 on the support of f. Note that the right-hand side does not depend on φ_0, since $f\varphi_0(x) = f$. From now on we define $[F,\psi(k)]$ by the formula

$$[F,\psi(k)] = [F,\psi(k)\varphi_0] = \int_{\mathbb{R}^3} f(x) f_\Psi(k,x)dx + 2\pi\alpha\gamma\bar{\nu}, \quad F \in \mathcal{H}^0_\alpha. \tag{3.6}$$

We formulate below our main Theorems and discuss their proofs in Sections 4 and 5.

Theorem 1. *There is an isometry V between $\mathcal{H}_\alpha$ and $L_2(\mathbb{R}^3)$ which is defined by the formulae*

$$(VF)(k) = [F,\psi(k)], \tag{3.7}$$

$$(V^{-1}\tilde{F})(x) = \int_{\mathbb{R}^3} \tilde{F}(k)\psi(k,x)dk, \tag{3.8}$$

for any $F \in \mathcal{H}^0_\alpha$ and $\tilde{F}$ in $C_0^\infty(\mathbb{R}^3)$.

Corollary 1. *Theorem 1 enables us to extend the definition of $[F,\psi(k)]$ to the whole space $\mathcal{H}_\alpha$ by the formula:*

$$[F,\psi(k)] = VF.$$

In fact, $[F, \psi(k)]$ is defined for $F : f \in C_0^\infty$ by formula (3.6) and according to the theorem $[F, \psi(k)] = VF$. Since the set $\mathcal{H}_\alpha^0$ is dense in $\mathcal{H}_\alpha$ and V is an isometry, we just extend the formula

$$[F, \psi(k)] = VF$$

to the whole space by continuity of V.

Corollary 2. *If $F(x)$ is such that $[F, \psi(k)] \in C_0(\mathbb{R}^3)$, then*

$$F(x) = \int_{\mathbb{R}^3} [F, \psi(k)]\psi(k, x)dk. \tag{3.9}$$

Let $\tilde{A}_\alpha$ be the operator of multiplication by $|k|^2$ in $L_2(\mathbb{R}^3)$, i.e.,

$$(\tilde{A}_\alpha \tilde{F})(k) = |k|^2 \tilde{F}(k),$$

$$D(\tilde{A}_\alpha) = \left\{ \tilde{F}(k) : \int_{\mathbb{R}^3} (\mid k \mid^4 + 1) \mid \tilde{F} \mid^2 dk < \infty \right\}.$$

Clearly, this operator is self-adjoint and its spectrum is $[0, \infty)$.

Let $A_\alpha = V^{-1} \tilde{A}_\alpha V$. By definition, the operator A_α is a self-adjoint operator in $\mathcal{H}_\alpha$.

$$D(A_\alpha) = \left\{ F : \int_{\mathbb{R}^3} (\mid k \mid^4 + 1) \mid [F, \psi(k)] \mid^2 dk < \infty \right\}. \tag{3.10}$$

$$A_\alpha F = \int_{\mathbb{R}^3} \mid k \mid^2 [F, \psi(k)]\psi(k, x)dk \tag{3.11}$$

for all $F : [F, \psi(k)] \in C_0(\mathbb{R}^3)$. It is enough to define the action of the operator on this set, since it is dense in the domain in the norm $\|A_\alpha \cdot \| + \| \cdot \|$.

Theorem 2. *Action of the operator A_α on its domain can be described by the formula:*

$$A_\alpha F(x) =_{x \neq 0} -\Delta F(x). \tag{3.12}$$

We will prove that $G'_{-\alpha^2}(x)$ can be represented as a linear combination of functions $\psi(k, x)$:

$$G'_{-\alpha^2}(x) = 2\pi\alpha \int_{\mathbb{R}^3} \overline{\nu(k)} \psi(k, x)dk, \tag{3.13}$$

$\nu(k)$ being given by formula (2.5). Function $G'_{-\alpha^2}(x)$ does not belong to the domain of A, since $|k|^2\nu(k)$ is not square integrable. Thus, $G'_{-\alpha^2}(x)$ is not an eigenfunction of $G'_{-\alpha^2}(x)$ even though it satisfies the equation

$$-\Delta G'_{-\alpha^2}(x) =_{x \neq 0} \alpha^2 G'_{-\alpha^2}(x).$$

Obviously, functions $\psi(k, x)$ form a complete set of generalized eigenfunctions of the operator A_α.

4. Proof of Theorem 1

The main goal is to construct the operator V satisfying Theorem 1. We prove the theorem in several steps. First, we introduce a set Q_α dense in $L_2(\mathbb{R}^3)$. We define $V' : Q_\alpha \to L_2(\mathbb{R}^3)$ by the formula $V'f = [f, \psi(k)]$ reduced to Q_α. We show that V' is isometric. We extend it by continuity to the whole space $L_2(\mathbb{R}^3)$. We prove that the formula $V'f = [f, \psi(k)]$ holds for all f with compact support. This part of the proof is covered by Lemmas 1–3. We denote the range of V' by H_0, $H_0 \subset L_2(\mathbb{R}^3)$. Introducing the operator $W = (V')^{-1}$ on H_0, we show that its action is described by (3.8). This is made in Lemmas 4–6. Further we prove that the co-dimension of H_0 in $L_2(\mathbb{R}^3)$ is equal to one. The subspace orthogonal to H_0 is spanned by the function $\overline{\nu(k)}$, where $\nu(k)$ is the scattering amplitude given by (2.5). This result is proven in Lemma 7. Next, we define $V : \mathcal{H}_\alpha \to L_2(\mathbb{R}^3)$ by the formula:

$$V \mid_{L_2(\mathbb{R}^3)} = V', \quad V'f = [f, \psi(k)], \tag{4.1}$$

$$VG'_{-\alpha^2}(x) = 2\pi\alpha\overline{\nu}(k). \tag{4.2}$$

By definition of $[\cdot, \psi(k)]$, see (3.6), $2\pi\alpha\overline{\nu}(k) = [G'_{-\alpha^2}, \psi(k)]$, since $f = 0$, $\gamma = 1$ for $F(x) = G'_{-\alpha^2}(x)$. Therefore,

$$VF = [F, \psi(k)] \quad \text{for any } F \in \mathcal{H}_\alpha. \tag{4.3}$$

Proving (3.13) (Lemma 8), we show that (3.8) holds on the orthogonal complement of H_0 too. Hence, (3.8) holds in the whole space $L_2(\mathbb{R}^3)$. Checking that $\|\overline{\nu}\|_{L_2(\mathbb{R}^3)} = (2\pi\alpha)^{-1/2}$ and considering the coefficient $2\pi\alpha$ in the definition (3.3) of $[\cdot, \cdot]$, we obtain that V is isometric (Lemma 9).

We start with introducing new notations.

1) Let $f \in C_0^\infty$, then

$$I_f(z) = \int_{\mathbb{R}^3} f(x)\overline{G}'_z(x)dx, \tag{4.4}$$

where,

$$G'_z(x) = \frac{\partial}{\partial x_1}\frac{\exp(-\sqrt{-z} \mid x \mid)}{\mid x \mid}, \quad Re\sqrt{-z} \geq 0, \tag{4.5}$$

(4.5) obviously being in agreement with (3.2).

2) Let

$$\tilde{f}(k) = \int_{\mathbb{R}_3} f(x)\overline{\psi}(k, x)dx, \tag{4.6}$$

$\psi(k, x)$ being given by (2.1), (2.5).

3) Let $\hat{f}(k)$ be the Fourier transform of f :

$$\hat{f}(k) = (2\pi)^{-3/2} \int_{R^3} f(x) \exp(-i(k, x))dx. \tag{4.7}$$

We prove some auxiliary lemmas.

Lemma 1. *If $f \in C_0^\infty$, then $(|k|^2 + 1)\tilde{f}(k) \in L_2(\mathbb{R}^3)$.*

Corollary 3. *If $f \in C_0^\infty$, then $\tilde{f}(k) \in L_1(\mathbb{R}^3)$.*

Proof. We prove the estimate:

$$\int_{\mathbb{R}^3} \mid \tilde{f} \mid^2 (\mid k \mid^2 + 1)^2 dk < \infty. \tag{4.8}$$

Obviously,

$$\tilde{f}(k) = \hat{f}(k) + \overline{\nu}(k) I_f(\mid k \mid^2 + i0), \tag{4.9}$$

where, in accordance with (4.4),

$$I_f(\mid k \mid^2 + i0) = \int_{\mathbb{R}^3} f(x) \frac{\partial}{\partial x_1} \left(\frac{\exp -i \mid k \mid\mid x \mid}{\mid x \mid} \right) dx. \tag{4.10}$$

Now we see that the integral (4.8) can be estimated from above by the sum:

$$2 \int_{\mathbb{R}^3} \mid \hat{f} \mid^2 (\mid k \mid^2 + 1)^2 dk + 2 \int_{\mathbb{R}^3} \mid \bar{\nu}(k) \mid^2 \mid I_f(\mid k \mid^2 + i0) \mid^2 (\mid k \mid^2 + 1)^2 dk. \tag{4.11}$$

The first integral converges since f is smooth. We prove the convergence of the second integral. It is easy to see from the definition (2.5) of $\nu(k)$ that $\bar{\nu}(k) \in L_2(\mathbb{R}^3)$. Therefore, it suffices to show that

$$|I_f(|k|^2 + i0)| < \frac{C}{|k|^2 + 1}. \tag{4.12}$$

We rewrite I_f in the form:

$$I_f(|k|^2 + i0) = \int_0^\infty \exp(-i \mid k \mid |x|)(g_1 - i \mid k \mid g_2) d|x|,$$

$$g_1(|x|) = -\int_{S_1} x_1 |x|^{-1} f(x) d\vartheta,$$

$$g_2(|x|) = \int_{S_1} x_1 f(x) d\vartheta,$$

S_1 being the unit sphere. Obviously, g_1, g_2 are twice continuously differentiable functions of $|x|$ with compact supports. Producing a Taylor expansion of $f(x)$ near the origin, we readily get: $g_1(0) = g_2(0) = g_2'(0) = 0$. Integration by parts yields (4.12). □

We denote by Q_α the set of functions $f \in C_0^\infty$ for which $I_f(-\alpha^2) = 0$. Note that $\overline{Q_\alpha} = L_2(\mathbb{R}^3)$, since $G'_{-\alpha^2} \notin L_2(\mathbb{R}^3)$. It is easy to see that

$$\tilde{f}(k) = [f, \psi(k)], \quad \text{when } f \in Q_\alpha, \tag{4.13}$$

$[f, \psi(k)]$ being given by (3.5) and (3.6).

Lemma 2. *If $f \in C_0^\infty$ and $g \in Q_\alpha$, then*

$$(\tilde{f}, \tilde{g}) = (f, g). \tag{4.14}$$

Proof. Since $(\hat{f},\hat{g}) = (f,g)$ it is suffices to prove that $I = 0$, I being defined by the formulae:

$$I = I_2 + I_2 + I_3, \tag{4.15}$$

$$I_1 = \int_{\mathbb{R}^3} \hat{f}(k)\nu(k)\overline{I}_g(|\, k\,|^2 + i0)dk,$$

$$I_2 = \int_{\mathbb{R}^3} \overline{\nu}(k)I_f(|\, k\,|^2 + i0)\overline{\hat{g}}(k)dk,$$

$$I_3 = \int_{\mathbb{R}^3} \nu(k)\overline{\nu}(k)I_f(|\, k\,|^2 + i0)\overline{I}_g(|\, k\,|^2 + i0)dk.$$

Note that the integrals I_1, I_2, I_3 converge, since $\hat{f}, \hat{g}, \nu \in L_2(\mathbb{R}^3)$ and I_f, I_g satisfy (4.12). We introduce the spherical coordinates $k = |\, k\,| \vartheta$ and integrate with respect to ϑ, taking into account that $I_f(|\, k\,|^2 \pm i0)$, $I_g(|\, k\,|^2 \pm i0)$ do not depend on ϑ and $\nu(k)$ depends on ϑ only through the factor k_1. For I_1 we obtain:

$$\int_{S_1} \hat{f}(k)\nu(k)d\vartheta = -3(2\pi)^{-3/2}(\alpha^3 + (i\,|\, k\,|)^3)^{-1}B(k), \tag{4.16}$$

where

$$B(k) = \int_{S_1} ik_1\hat{f}(k)d\vartheta. \tag{4.17}$$

We show that

$$B(k) = -i(2\pi)^{1/2}|k|^{-1}(I_f(|\, k\,|^2 + i0) - I_f(|\, k\,|^2 - i0)), \tag{4.18}$$

where

$$I_f(|\, k\,|^2 - i0) = \int_{\mathbb{R}^3} f(x)\frac{\partial}{\partial x_1}\left(\frac{\exp i\,|\, k\,||\, x\,|}{|\, x\,|}\right)dx. \tag{4.19}$$

In fact,

$$\begin{aligned} B(k) &= \int_{S_1} ik_1\hat{f}(k)d\vartheta \\ &= (2\pi)^{-3/2}\int_{S_1} ik_1 \int_{\mathbb{R}^3} f(x)\exp -i(k,x)dxd\vartheta \\ &= -(2\pi)^{-3/2}\int_{\mathbb{R}^3} f(x)\frac{\partial}{\partial x_1}\int_{S_1} \exp -i(k,x)d\vartheta. \end{aligned} \tag{4.20}$$

Using the well-known relation

$$\frac{i|k|}{2\pi}\int_{S_1} \exp -i(k,x)d\vartheta = \frac{\exp(i\,|\, k\,||\, x\,|)}{|\, x\,|} - \frac{\exp(-i\,|\, k\,||\, x\,|)}{|\, x\,|}, \tag{4.21}$$

we obtain:

$$\begin{aligned} B(k) &= i(2\pi)^{-1/2}|k|^{-1}\int_{\mathbb{R}^3} f(x)(G'_{|k|^2+i0}(x) - G'_{|k|^2-i0}(x))dx \\ &= i(2\pi)^{-1/2}|k|^{-1}\left(I_f(|\, k\,|^2 - i0) - I_f(|\, k\,|^2 + i0)\right). \end{aligned} \tag{4.22}$$

Thus, (4.18) is proved. Substituting (4.16) and (4.18) in the integral I_1, we get:

$$I_1 = 3i(2\pi)^{-2}\int_0^\infty \frac{\left(I_f(\mid k\mid^2 + i0) - I_f(\mid k\mid^2 - i0)\right)\overline{I}_g(\mid k\mid^2 + i0)}{\alpha^3 + (i\mid k\mid)^3}\mid k\mid d\mid k\mid . \tag{4.23}$$

Obviously, $I_2(f,g) = \overline{I_1(g,f)}$. Therefore,

$$I_2 = -3i(2\pi)^{-2}\int_0^\infty \frac{\left(\overline{I}_g(\mid k\mid^2 + i0) - \overline{I}_g(\mid k\mid^2 - i0)\right)I_f(\mid k\mid^2 + i0)}{\alpha^3 - (i\mid k\mid)^3}\mid k\mid d\mid k\mid . \tag{4.24}$$

In I_3 the integration is trivial:

$$\begin{aligned}\int_{S_1}\nu(k)\overline{\nu}(k)d\vartheta &= 9(2\pi)^{-3}\int_{S_1}\frac{k_1^2}{(\alpha^3 - (i\mid k\mid)^3)(\alpha^3 + (i\mid k\mid)^3}d\vartheta \\ &= \frac{6(2\pi)^{-2}\mid k\mid^2}{(\alpha^3 - (i\mid k\mid)^3)(\alpha^3 + (i\mid k\mid)^3)}.\end{aligned} \tag{4.25}$$

Thus

$$I_3 = 6(2\pi)^{-2}\int_0^\infty \frac{I_f(\mid k\mid^2 + i0)\overline{I}_g(\mid k\mid^2 + i0)}{(\alpha^3 - (i\mid k\mid)^3)(\alpha^3 + (i\mid k\mid)^3}\mid k\mid^4 d|k|. \tag{4.26}$$

Note that

$$-3(2\pi)^{-2}i\left(\frac{|k|}{\alpha^3 - (i|k|)^3} - \frac{|k|}{\alpha^3 + (i|k|)^3}\right) + \frac{6(2\pi)^{-2}|k|^4}{(\alpha^3 - (i|k|)^3)(\alpha^3 + (i|k|)^3)} = 0. \tag{4.27}$$

Adding (4.23), (4.24), (4.26) and using the last relation we obtain:

$$\begin{aligned}I = &-3i(2\pi)^{-2}\int_0^\infty \frac{I_f(\mid k\mid^2 - i0)\overline{I}_g(\mid k\mid^2 + i0)}{\alpha^3 + (i\mid k\mid)^3}\mid k\mid d|k| \\ &+ 3i(2\pi)^{-2}\int_0^\infty \frac{I_f(\mid k\mid^2 + i0)\overline{I}_g(\mid k\mid^2 - i0)}{\alpha^3 - (i\mid k\mid)^3}\mid k\mid d|k|.\end{aligned} \tag{4.28}$$

The integrals converge, since I_f, I_g are bounded when $k \to \infty$. Making the change of the variables $|k| \to -|k|$ in the second term, we arrive to the formula:

$$I = -3i(2\pi)^{-2}\int_{-\infty}^\infty \frac{I_f(\mid k\mid^2 - i0)\overline{I}_g(\mid k\mid^2 + i0)}{\alpha^3 + (i\mid k\mid)^3}\mid k\mid dk. \tag{4.29}$$

The integrand is an analytical exponentially decaying function of $\mid k\mid$ in the upper half-plane, since the function (4.19) depends on $\mid k\mid$ analytically in the upper half-plane decaying at infinity and $\overline{I}_g(k^2 + i0) = I_g(k^2 - i0)$. The integrand has a single pole in the upper half-plane at the point $\mid k\mid = i\alpha$. Calculating the integral by residue calculus, we obtain:

$$I = \frac{1}{2\pi\alpha}I_f(-\alpha^2)\overline{I_g(-\alpha^2)}. \tag{4.30}$$

Since $g \in Q_\alpha$, we have $I_g(-\alpha^2) = 0$. Therefore, $I = 0$. □

Next, we introduce the operator $V' : L_2(\mathbb{R}^3) \to L_2(\mathbb{R}^3)$ by the formulae:

$$D(V') = Q_\alpha, \tag{4.31}$$

$$V'f(x) = [f, \psi(k)]. \tag{4.32}$$

By (4.13) and Lemma 2, V' is an isometry:

$$\|V'f\|_{L_2(\mathbb{R}^3)} = \|f\|_{L_2(\mathbb{R}^3)}. \tag{4.33}$$

Since $\overline{Q_\alpha} = L_2(\mathbb{R}^3)$, V' can be extended by continuity to the whole space $L_2(\mathbb{R}^3)$. Thus, the operator V' acts from $L_2(\mathbb{R}^3)$ on a subspace of $L_2(\mathbb{R}^3)$, which we denote by H_0, i.e.,

$$H_0 = \overline{V'Q_\alpha}. \tag{4.34}$$

Lemma 3. *If $f \in L_2(\mathbb{R}^3)$ and has a compact support, then*

$$(V'f)(k) = \int_{\mathbb{R}^3} f(x)\overline{\Big(\psi(k,x) - \nu(k)G'_{-\alpha^2}(x)\Big)}dx. \tag{4.35}$$

Corollary 4. *If $f \in L^2(\mathbb{R}^3)$ and has a compact support, then*

$$V'f = [f, \psi(k)]. \tag{4.36}$$

We get the corollary comparing (4.35) with (3.5), (3.6), where $\gamma = 0$.

Proof. In fact, let $f_n \in Q_\alpha$, $n \in \mathbf{N}$, and $\lim_{n\to\infty} f_n = f$ in $L_2(\mathbb{R}^3)$, the support of functions f_n being uniformly bounded. It is easy to see that

$$\begin{aligned}(V'f_n)(k) &= \int_{\mathbb{R}^3} f_n(x)\overline{\psi(k,x)}dx \\ &= \int_{\mathbb{R}^3} f_n(x)\overline{\psi(k,x) - \nu(k)G'_{-\alpha^2}(x)}dx.\end{aligned} \tag{4.37}$$

The function $\psi(k,x) - \nu G'_{-\alpha^2}(x)$ is in $L_{2,\mathrm{loc}}$, since it has a singularity at zero of the type $O(\frac{1}{|x|})$. Therefore, we can pass to the limit in the last integral for any k. □

We use formula (4.36) to extend the definition of $[f, \psi(k)]$ to the whole space $L_2(\mathbb{R}^3)$. Namely, for any $f \in L_2(\mathbb{R}^3)$

$$[f, \psi(k)] = (V'f)(k). \tag{4.38}$$

Next, let us consider $\phi(k) : \sqrt{|k|^2+1}\phi(k) \in L_2(\mathbb{R}^3)$ and the integral:

$$(\mathcal{W}\phi)(x) = \int_{\mathbb{R}^3} \phi(k)\psi(k,x)dk. \tag{4.39}$$

Lemma 4. *If $\phi(k) : \sqrt{|k|^2+1}\phi(k) \in L_2(\mathbb{R}^3)$, then $(\mathcal{W}\phi)(x) \in L_{1,\mathrm{loc}}(\mathbb{R}^3)$.*

Proof. To make sense out of the right part, we first assume that $\phi(k) \in C_0^\infty(\mathbb{R}^3)$. Obviously, the integral converges for all $x \neq 0$ and

$$\begin{aligned}(\mathcal{W}\phi)(x) \quad &= \quad \frac{1}{(2\pi)^{3/2}}\int_{\mathbb{R}^3} \phi(k)\exp i(k,x)dk \\ &+ \int_{\mathbb{R}^3} \phi(k)\nu(k)G'(|k|^2+i0)(x)dk.\end{aligned} \tag{4.40}$$

The first integral is just the Fourier transform of $\phi(k)$ and, therefore, can be extended by continuity to all function in $L_2(\mathbb{R}^3)$. The second integral converges for all $\phi(k) : \sqrt{|k|^2+1}\phi(k) \in L_2$, since $\nu(k) \in L_2(\mathbb{R}^3)$ and $G'(|k|^2+i0)(x)$ satisfies the inequality:

$$|G'_{|k|^2+i0}(x)| < \frac{c\sqrt{|k|^2+1}\sqrt{|x|^2+1}}{|x|^2}, \tag{4.41}$$

when $x \neq 0$. Thus, formula (4.40) defines $(W\phi)(x)$ for all $\phi(k) : \sqrt{|k|^2+1}\phi(k) \in L_2(\mathbb{R}^3)$. It is easy to see that $(W\phi)(x) \in L_{1,\mathrm{loc}}(\mathbb{R}^3)$. In fact, the first term on the r.h.s. of (4.40) is a square-integrable function of x, and the second term is a continuous function of x when $x \neq 0$ and has a singularity of the type $O(\frac{1}{|x|^2})$ at $x = 0$. □

Lemma 5. *If* $f(x) \in Q_\alpha$, *then*

$$f(x) = (\mathcal{W}\tilde{f})(x). \tag{4.42}$$

$\tilde{f}$, $\mathcal{W}$ being defined by (4.6), (4.39).

Proof. Let $g(x) \in C_0^\infty(\mathbb{R}^3)$. Using (4.39) and formally exchanging the order of integration with respect to x and k, we obtain:

$$\int_{\mathbb{R}^3} (W\tilde{f})(x)\overline{g(x)}dx = \int_{\mathbb{R}^3} \tilde{f}(k)\overline{\tilde{g}(k)}dk = (\tilde{f}, \tilde{g}). \tag{4.43}$$

By Lemma 2, $(\tilde{f}, \tilde{g}) = (f, g)$. Therefore,

$$\int_{\mathbb{R}^3} (W\tilde{f})(x)\overline{g(x)}dx = \int_{\mathbb{R}^3} f(x)\overline{g(x)}dx.$$

The last relation holds for any $g(x) \in C_0^\infty(\mathbb{R}^3)$, and $(W\tilde{f})(x) \in L_{1,\mathrm{loc}}(\mathbb{R}^3)$. Therefore, (4.42) is true. It remains to justify the change of the order of integration. It is enough to show that

$$\tilde{f}(k)\psi(k,x)\bar{g}(x) \in L_1(\mathbb{R}^3 \times \mathbb{R}^3) \tag{4.44}$$

and to apply Fubini's theorem. We consider $\tilde{f}(k)\psi(k,x)\bar{g}(x)$ as a sum of two functions:

$$K_1(k,x) = \tilde{f}(k)e^{i(k,x)}\bar{g}(x), \qquad K_2(k,x) = \tilde{f}(k)\nu(k)G'_{|k|^2+i0}(x)\bar{g}(x).$$

By Corollary 3, $\tilde{f} \in L_1(\mathbb{R}^3)$. Hence, $K_1(k,x) \in L_1(\mathbb{R}^3 \times \mathbb{R}^3)$. Using Lemma 1, the estimate (4.41) and the relation $\nu(k) \in L_2(\mathbb{R}^3)$, it is easy to show that $K_2(k,x) \in L_1(\mathbb{R}^3 \times \mathbb{R}^3)$. Thus, (4.44) holds. □

Let us define the operator $W : H_0 \to L_2(\mathbb{R}^3)$ as follows. Its domain $D(W)$ is $V'Q_\alpha$, see (4.31), (4.32). For any $\phi \in D(W)$,

$$W\phi = \mathcal{W}\phi. \tag{4.45}$$

By Lemmas 2 and 5 the operator W is isometric. Therefore, it can be extended by continuity to the whole subspace H_0.

Let I be the identity operator in $L_2(\mathbb{R}^3)$ and I_0 be the identity operator in H_0.

Lemma 6. *The operators V', W satisfy to the following relations:*

$$WV' = I. \tag{4.46}$$

$$V'W = I_0, \tag{4.47}$$

Proof. By Lemma 5 and formula (4.32), $WV'f = f$ for any $f \in Q_\alpha$. Since V' and W are continuous, it is true for any f in $L_2(\mathbb{R}^3)$, i.e., (4.46) holds. Applying W to the both parts of (4.46) and considering (4.32), we obtain: $V'W\tilde{f} = \tilde{f}$ for any $\tilde{f} \in V'Q_\alpha$. Since V' and W are continuous, it is true for any $\tilde{f}$ in H_0, i.e., (4.47) holds. □

Now we consider the subspace H_0, i.e., the set of functions, which can be represented in the form:

$$F(k) = [f, \psi(k)], \quad f \in L_2(\mathbb{R}^3).$$

We determine the co-dimension of H_0.

Lemma 7. *The co-dimension of the subspace H_0 in $L_2(\mathbb{R}^3)$ is equal to* 1. *The function $\overline{\nu(k)}$ is orthogonal to H_0 in $L_2(\mathbb{R}^3)$. Formula* (4.45) *holds for all $\phi \in H_0 \cap C_0^\infty$.*

Proof. Let us denote by H_0' the subspace of $L_2(\mathbb{R}^3)$ orthogonal to $\bar{\nu}(k)$. The goal is to prove that $H_0' = H_0$.

First we prove that $H_0 \subset H_0'$. Assume that $f \in Q_\alpha$. We prove that the function $[f, \psi(k)]$ is orthogonal $\overline{\nu}(k)$ in $L_2(\mathbb{R}^3)$, that is $\gamma_f = 0$, γ_f being the integral:

$$\gamma_f = \int_{\mathbb{R}^3} [f, \psi(k)]\nu(k)dk. \tag{4.48}$$

According to Lemma 5,

$$f(x) = \int_{\mathbb{R}^3} [f, \psi(k)]\psi(k, x)dk. \tag{4.49}$$

Therefore,

$$f(x) = F_1 + F_2 + \gamma_f G'_{-\alpha^2}(x), \tag{4.50}$$

where

$$F_1 = (2\pi)^{-3/2} \int_{\mathbb{R}^3} [f, \psi(k)] \exp i(k, x)dk, \tag{4.51}$$

$$F_2 = \int_{\mathbb{R}^3} [f, \psi(k)]\nu(k)(G'_{k^2+i0}(x) - G'_{-\alpha^2}(x))dk. \tag{4.52}$$

Suppose we have proved the relations

$$F_1 \in L_{2,\mathrm{loc}}(\mathbb{R}^3), \quad F_2 \in L_{2,\mathrm{loc}}(\mathbb{R}^3). \tag{4.53}$$

Then, taking into account that f is a smooth function, we obtain $\gamma_f = 0$. Thus, it suffices to check (4.53). It is easy to see that $F_1 \in L_2(\mathbb{R}^3)$, since $[f, \psi(k)] \in H_0 \subset L_2(\mathbb{R}^3)$. Next, we use the obvious inequality

$$| G'_{k^2+i0}(x) - G'_{-\alpha^2}(x) |< c\sqrt{|k|^2+1}\,|x|^{-1},\ c \neq c(k).$$

Moreover, $\sqrt{|k|^2+1}\,[f, \psi(k)] \in L_2(\mathbb{R}^3)$ by Lemma 1. Taking into account also that $\nu(k) \in L_2(\mathbb{R}^3)$, we obtain $F_2 \in L_{2,\text{loc}}$. Thus, we have shown that $\gamma_f = 0$, when $f \in Q_\alpha$. This means that $V'Q_\alpha$ is orthogonal to $\bar{\nu}(k)$. Taking into account (4.32), (4.34), (4.38), we obtain that H_0 is orthogonal to $\bar{\nu}(k)$, i.e., $H_0 \subset H'_0$.

Next we prove that $H'_0 \subset H_0$. The set of smooth functions, vanishing at infinity, is dense in H'_0. Therefore, it suffice to prove that any function $\phi(k) \in C_0^\infty(\mathbb{R}^3)$ orthogonal to $\overline{\nu(k)}$ belongs to H_0. We start with considering $(\mathcal{W}\phi)(k)$, see (4.39). By Lemma 4, $(\mathcal{W}\phi)(k) \in L_{1,\text{loc}}$. Suppose we have proved that

$$\|\mathcal{W}\phi\|_{L_2(\mathbb{R}^3)} = \|\phi\|_{L_2(\mathbb{R}^3)},\ \text{when } \phi \in H'_0 \cap C_0^\infty. \tag{4.54}$$

Then, we define the operator $W' : H'_0 \to L_2(\mathbb{R}^3)$ as follows:

$$D(W') = H'_0 \cap C_0^\infty,\quad W'f = \mathcal{W}f. \tag{4.55}$$

Formula (4.54) means that W' is isometric and therefore can be extended by continuity to the whole subspace H'_0. Considering that $H_0 \subset H'_0$ and the definitions of W, W', we obtain that $W \subset W'$.

Let

$$f(x) = (W'\phi)(x),\quad \phi \in H'_0 \cap C_0^\infty.$$

By Lemma 6,

$$f = WV'f = W'V'f,$$

since $V'f \in H_0$ by the definition of H_0. By the first part of this lemma $V'f \in H'_0$. Since W' is isometry and $W'\phi = W'V'f$,

$$\phi(k) = V'f,$$

and therefore, $\phi \in H_0$. The last relation holds for any $\phi \in H'_0$, therefore, $H'_0 \subset H_0$. Thus, $H_0 = H'_0$ and $W = W'$. Formula (4.45) follows from (4.55) and the latter relation. Thus, it remains to prove (4.54).

The Fourier transform of $f(x) = \mathcal{W}\phi$ is obviously given by the formula:

$$\hat{f}(\xi) = \phi(\xi) + \int_{\mathbb{R}^3} \phi(k)\nu(k)\varphi_1(\xi, k)dk, \tag{4.56}$$

where $\varphi_1(k, \tau)$ is the Fourier transform of the function $G'_{|k|^2+i0}(x)$:

$$\varphi_1(\xi, k) = \frac{2(2\pi)^{-1/2} i\xi_1}{|\ \xi\ |^2 - |\ k\ |^2 - i0}.$$

It is not difficult to check that $\hat{f}(\xi)$ is continuous. It is clear that

$$\|f\|_{L_2(\mathbb{R}^3)} = \|\hat{f}\|_{L_2(\mathbb{R}^3)} = I_0 + I_1 + I_2 + I_3, \tag{4.57}$$

$$I_0 = \lim_{R\to\infty} \int_{|\xi|<R} |\phi(\xi)|^2 d\xi,$$

$$I_1 = \lim_{R\to\infty} \int_{|\xi|<R} \int_{\mathbb{R}^3} \phi(k)\overline{\phi}(\xi)\nu(k)\varphi_1(\xi,k)dkd\xi,$$

$$I_2 = \lim_{R\to\infty} \int_{|\xi|<R} \int_{\mathbb{R}^3} \phi(\xi)\overline{\phi}(\tau)\overline{\nu}(\tau)\overline{\varphi_1}(\xi,\tau)d\tau d\xi,$$

$$I_3 = \lim_{R\to\infty} \int_{\mathbb{R}^3\times\mathbb{R}^3} \phi(k)\overline{\phi}(\tau)\nu(k)\overline{\nu}(\tau)\varphi_R(k,\tau)dkd\tau,$$

where

$$\varphi_R(k,\tau) = \int_{|\xi|<R} \varphi_1(\xi,k)\overline{\varphi}_1(\xi,\tau)d\xi.$$

Obviously, $I_0 = \|\phi\|_{L_2(\mathbb{R}^3)}$. Since $\phi(\xi) \in C_0^\infty(\mathbb{R}^3)$, the passage to the limit is trivial in I_1 and I_2. Changing the names of variables of integration for more convenient ones (the same for both integrals), we obtain:

$$I_1 = \int_{\mathbb{R}^3\times\mathbb{R}^3} \phi(k)\overline{\phi}(\tau)\nu(k)\varphi_1(\tau,k)dkd\tau,$$

$$I_2 = \int_{\mathbb{R}^3\times\mathbb{R}^3} \phi(k)\overline{\phi}(\tau)\overline{\nu}(\tau)\overline{\varphi_1}(k,\tau)dkd\tau.$$

A straightforward calculation gives:

$$\nu(k)\varphi_1(\tau,k) + \overline{\nu}(\tau)\overline{\varphi_1}(k,\tau) = -\frac{i4\pi\nu(k)\overline{\nu(\tau)}(|\tau|^3 + |k|^3)}{3(|k|^2 - |\tau|^2 + i0)}.$$

Therefore,

$$I_1 + I_2 = -\int_{\mathbb{R}^3\times\mathbb{R}^3} \frac{4\pi i\phi(k)\overline{\phi}(\tau)\nu(k)\overline{\nu(\tau)}(|\tau|^3 + |k|^3)}{3(|k|^2 - |\tau|^2 + i0)} dkd\tau.$$

Next we calculate φ_R. Integrating over the unit sphere with respect to ξ we obtain:

$$\begin{aligned}\varphi_R &= 8/3 \int_0^R \frac{r^4}{(r^2 - |k|^2 - i0)(r^2 - |\tau|^2 + i0)} dr \\ &= 4/3 \int_{-\infty}^{\infty} \left(\frac{r^4}{(r^2 - |k|^2 - i0)(r^2 - |\tau|^2 + i0)} - 1 \right) dr + 8R/3 + O(R^{-1}).\end{aligned}$$

The integral on the right-hand side can easily be calculated using residues at the points $r = |k| + i0$ and $r = -|\tau| + i0$. Thus,

$$\varphi_R = \frac{4\pi i(|\tau|^3 + |k|^3)}{3(|k|^2 - |\tau|^2 + i0)} + \frac{8\pi R}{3} + o(R).$$

Substituting φ_R into the formula for I_3 and taking into account, that $\phi(k)$ is orthogonal to $\overline{\nu}(k)$, we get that the part of the integral containing $8R/3$ vanishes and, therefore,

$$\begin{aligned} I_3 &= 4/3\pi i \int_{\mathbb{R}^3\times\mathbb{R}^3} \frac{4\pi i\phi(k)\phi(\tau)\nu(k)\overline{\nu(\tau)}(\mid \tau \mid^3 + \mid k \mid^3)}{3(\mid k \mid^2 - \mid \tau \mid^2 + i0)} dk d\tau \\ &= -(I_1 + I_2). \end{aligned}$$

Using the relation $I_3 = -(I_1 + I_2)$ and formula (4.57) we obtain (4.54). □

Next, we consider the function $\overline{\nu}(k)$ being orthogonal to H_0. Taking into account that $\int_0^\infty \frac{r^4\, dr}{1+r^6} = \frac{\pi}{3}$, it is not hard to calculate the norm of $\overline{\nu}(k)$ in $L_2(\mathbb{R}^3)$:

$$\|\overline{\nu}\| \equiv \|\overline{\nu}\|_{L_2(\mathbb{R}^3)} = \left(\int_{\mathbb{R}^3} \frac{9k_1^2}{(2\pi)^3(\alpha^6 + \mid k \mid^6)} dk\right)^{1/2} = (2\pi\alpha)^{-1/2}. \tag{4.58}$$

Now we determine the function, which corresponds to $\overline{\nu}(k)$ in the x-representation.

Lemma 8. *The following formula for $G'_{-\alpha^2}(x)$ is valid:*

$$G'_{-\alpha^2}(x) = 2\pi\alpha \int_{\mathbb{R}^3} \overline{\nu(k)}\psi(k,x)dk. \tag{4.59}$$

Proof. To calculate the integral on the right-hand side we represent it as a sum of A_1 and A_2:

$$\begin{aligned} A_1 &= (2\pi)^{-3/2} \int_{\mathbb{R}^3} \overline{\nu(k)} \exp i(k,x) dk, \\ A_2 &= \int_{\mathbb{R}^3} |\nu(k)|^2 G'_{|k|^2+i0}(x) dk. \end{aligned}$$

Using relation (4.21), we obtain:

$$A_1 = \int_0^\infty \frac{3|k| \left(G'_{|k|^2+i0} - G'_{|k|^2-i0}\right)(x)}{i(2\pi)^2(\alpha^3 - (i|k|)^3)} dk.$$

Integrating with respect to angle variables, we get:

$$A_2 = \int_0^\infty \frac{3 \mid k \mid^4}{2\pi^2(\alpha^6 + \mid k \mid^6)} G'_{|k|^2+i0}(x) d|k|.$$

Taking into account the obvious relation

$$\frac{2 \mid k \mid^4}{\alpha^6 + \mid k \mid^6} = \frac{i|k|}{\alpha^3 - (i \mid k \mid)^3} - \frac{i|k|}{\alpha^3 + (i \mid k \mid)^3},$$

we see that

$$4\pi^2(A_1 + A_2) = \int_0^{\infty} \frac{3i \mid k \mid}{\alpha^3 - (i \mid k \mid)^3} G'_{|k|^2 - i0(x)} d \mid k \mid$$
$$- \int_0^{\infty} \frac{3i \mid k \mid}{\alpha^3 + (i \mid k \mid)^3} G'_{|k|^2 + i0}(x) d \mid k \mid$$
$$= - \int_{-\infty}^{\infty} \frac{3it}{\alpha^3 + (it)^3} \frac{\partial}{\partial x_1} \frac{e^{it|x|}}{|x|} dt.$$

The integrand on the right-hand side has a single pole in the upper half-plane and rapidly decays, when t tends to infinity. Calculating the residue at the point $t = i\alpha$, we obtain:

$$A_1 + A_2 = (2\pi\alpha)^{-1} G'_{-\alpha^2}(x).$$ □

We consider the operator $V : \mathcal{H}_\alpha \to L_2(\mathbb{R}^3)$, defined as follows:

$$V \mid_{L_2(\mathbb{R}^3)} = V', \tag{4.60}$$

$$VG'_{-\alpha^2}(x) = 2\pi\alpha\overline{\nu}(k). \tag{4.61}$$

By definition (3.6) of $[\cdot, \psi(k)]$,

$$[G'_{-\alpha^2}, \psi(k)] = 2\pi\alpha\overline{\nu}(k), \tag{4.62}$$

since $f = 0$, $\gamma = 1$, when $F(x) = G'_{-\alpha^2}(x)$. Using also (4.36), we obtain:

$$VF = [F, \psi(k)], \text{ for all } F \in \mathcal{H}_\alpha. \tag{4.63}$$

Lemma 9. *Operator V is an isometric isomorphism between $\mathcal{H}_\alpha$ and $L_2(\mathbb{R}^3)$.*

Proof. Since operator V' is defined on the whole space $L_2(\mathbb{R}^3)$, formulas (4.60), (4.61) define operator V on $\mathcal{H}_\alpha$. We prove that

$$\|VF\|_{L_2(\mathbb{R}^3)} = \|F\|_{H_\alpha}, \tag{4.64}$$

for any F in $\mathcal{H}_\alpha$. Using formulae (3.1), (4.60) and (4.61) we obtain:

$$VF = V'f + \gamma 2\pi\alpha\overline{\nu}(k). \tag{4.65}$$

Taking into account, that $V'f \in H_0$ and $\overline{\nu}(k)$ is orthogonal to H_0 (see Lemma 7), we get:

$$\|VF\|^2_{L_2(\mathbb{R}^3)} = \|V'f\|^2_{L_2(\mathbb{R}^3)} + \mid \gamma \mid^2 4\pi^2\alpha^2 \|\overline{\nu}\|^2_{L_2(\mathbb{R}^3)}.$$

Taking into account that V' is an isometry and using relations (4.58) and (3.3), we obtain (4.64).

According to Lemma 7 the element $\overline{\nu}(k)$ spans up the orthogonal complement of H_0 in $L_2(\mathbb{R}^3)$. By the definition of H_0 we have $V'L_2(\mathbb{R}^3) = H_0$. By the definition of V we have $VG'_{-\alpha^2}(x) = 2\pi\alpha\overline{\nu}(k)$, i.e., $VG'_{-\alpha^2}$ is orthogonal to H_0. Therefore, the range of V is $L_2(\mathbb{R}^3)$. □

Proof of Theorem 1. By Lemma 9, V is an isometry. Formula (3.7) is the same as (4.63). To prove relation (3.8) we represent $\tilde{F} \in L_2(\mathbb{R}^3)$ in the form:

$$\tilde{F} = \phi + \tilde{\gamma}\overline{\nu}(k), \quad \phi \in H_0. \tag{4.66}$$

According to the definition of operator V

$$V^{-1}\phi = V'^{-1}\phi,$$
$$V^{-1}\overline{\nu}(k) = (2\pi\alpha)^{-1}G'_{-\alpha^2}(x)$$

By Lemma 6, $V'^{-1}\phi = W\phi$. Suppose $\phi \in H_0 \cap C_0^\infty$. By Lemma 7,

$$W\phi = \int_{\mathbb{R}^3} \phi(k)\psi(k,x)dk. \tag{4.67}$$

Using relations (4.59) and (4.61) we obtain

$$\tilde{\gamma}V^{-1}\overline{\nu}(k) = \int_{\mathbb{R}^3} \tilde{\gamma}\overline{\nu}(k)\psi(k,x)dk.$$

Adding the last two equalities and using formula (4.66), we verify relation (3.8). □

5. Construction of the energy operator

We consider in $L_2(\mathbb{R}^3)$ the operator of multiplication by $\mid k \mid^2$:

$$\tilde{A}_\alpha \tilde{F}(k) = \mid k \mid^2 \tilde{F}(k) \tag{5.1}$$

Naturally, its range is the set of functions, for which the integral

$$\int_{\mathbb{R}^3} (\mid k \mid^4 + 1) \mid \tilde{F}(k) \mid^2 dk \tag{5.2}$$

converges.

Next, we consider the operator $A_\alpha = V^{-1}\tilde{A}_\alpha V$. Using Theorem 1 we obtain that the domain of A is described by the condition:

$$\int_{\mathbb{R}^3} (\mid k \mid^4 + 1) \mid [F, \psi(k)] \mid^2 dk < \infty \tag{5.3}$$

The action of the operator A_α is described by the formula

$$A_\alpha F(x) = \int_{\mathbb{R}^3} \mid k \mid^2 [F, \psi(k)]\psi(k,x)dk \tag{5.4}$$

for functions $F(x)$ such that $[F, \psi(k)] \in C_0(\mathbb{R}^3)$. For other $F(x)$ in $D(A_\alpha)$ we can obtain $A_\alpha F$ by closing (5.4) in the norm of $\| \cdot \| + \|A_\alpha \cdot \|$.

Proof of Theorem 2. Suppose F is such that $[F, \psi(k)] \in C_0^\infty(\mathbb{R}^3)$. It is clear that $F \in D(A_\alpha)$, since the condition (5.3) is satisfied. The set $C_0^\infty(\mathbb{R}^3)$ is dense in the domain of A_α with respect to the norm $\| \cdot \| + \|A_\alpha \cdot \|$. Therefore, it suffices to verify formula (3.12) assuming that $[F, \psi(k)]$ is a function with compact support. Relation (3.12) is obvious, because $\mid k \mid^2 \psi(k,x) =_{x\neq 0} -\Delta_x\psi(k,x)$. □

Remark. Note that $G'_{-\alpha^2}(x)$ does not belong to $D(A)$. In fact,

$$[G'_{-\alpha^2}(x), \psi(k)] = \overline{\nu}(k) \tag{5.5}$$

But $\overline{\nu}(k) \notin D(A_\alpha)$, since the integral (5.2) diverges for $\tilde{F} = \overline{\nu}$. By Lemma 8, $G'_{-\alpha^2}(x)$ can be represented as an integral of functions $\psi(k,x)$. It follows that $G'_{-\alpha^2}(x)$ is not an eigenfunction of A_α, even though it formally satisfies the differential equation for an eigenfunction.

6. Possible Generalizations

We can consider the case of spherically symmetric inclusion by taking P-scattered waves in the form:

$$\psi(k,x) = (2\pi)^{-3/2} \exp i(k,x) + \sum_{j=1}^{3} \nu_j(k) \frac{\partial G_{|k|^2+i0}(x)}{\partial x_j}, \tag{6.1}$$

$$\nu_j(k) = -\frac{(2\pi)^{-3/2} 3ik_j}{\alpha^3 + (i \mid k \mid)^3}. \tag{6.2}$$

It correspond to a spherically symmetric inclusion, since scattered waves depend on x only through the product (k,x). All considerations are analogous to the previous case.

It is not hard to generalize all theorems to the case of negative α, taking instead of function $G'_{-\alpha^2}(x)$ the function

$$G'_{-\alpha_+^2}(x) + iG'_{-\alpha_-^2}(x), \alpha_\pm = \mid \alpha \mid \exp \pm i\pi/3.$$

In formula (3.3) instead of α one should substitute $-\alpha$.

The model can be generalized also to the case of stronger singularities, of the types

$$\frac{\partial^{|m|}}{\partial x_1^{m_1} \partial x_2 m_2 \partial x_3^{m_3}} \frac{e^{-\alpha}}{|x|},$$

which correspond to scattered waves of higher orders. In the case $\mid m \mid \neq 1$ one can obtain $H_{\vec{\alpha}}$ by adding $N, N = [m/2] + 1$, singular elements to $L_2(\mathbb{R}^3)$.

Finally, it is possible to generalize the model to the case of α which are functions of $|k|$: $\alpha = \alpha(|k|)$. In particular, Theorem 1 is valid, when

1. Function $\alpha(|k|)$ is analytic in the upper half-plane with respect to $\mid k \mid$.
2. Function $\alpha(k)$ is real and continuous for real $\mid k \mid$.
3. $\nu(k) =_{|k| \to \infty} O(|k|^{-2})$, where $\nu(k) = 3ik_1(\alpha(k)^3 + (i \mid k \mid)^3)^{-1}$.
4. Function $\alpha(|k|)^3 + (i \mid k \mid)^3$ has an unique zero in the upper half-plane. This zero belongs to the imagine axis.
5. The function $(\mid \alpha(k)^3 - (i \mid k \mid)^3)^{-1}$ has a subexponential growth when $|k| = Re^{i\vartheta}$, $R \to \infty$, $0 < \vartheta < \pi$.

References

[1] E. Fermi, *Sul moto dei neutroni nelle sostanzeidrogenate*, Ricerca Scientifica, **7** (1936), 13–52, English translation in E. Fermi, Collected Papers, vol. I, Italy 1921–1938, University of Chicago Press, Chicago, 1962, pp. 980–1016.

[2] A.I. Baz, Ya.B. Zeldovich, and A.M. Perelomov, *Scattering,reactions, and decay in nonrelativistic Quantum mechanics* (in Russian), Moscow, 1971.

[3] Yu.N. Demkov and V.N. Ostrovsky, *The method of potentials of zero radius in atomic physics* (in Russian), Leningrad, 1975.

[4] L.C. Evans, *Partial differential equations*, in the series Graduate Study in Mathematics, Vol. 19, AMS, 1998.

[5] F.A. Berezin and L.D. Faddeev, *Remark on the Schrödinger equation with singular potential*, Dokl. Acad. Nauk SSSR, **137** (1961), 1011–1014.

[6] N.I. Akhiezer and I.M. Glazman, *Theory of Linear Operators in Hilbert Space*, Dover Books, 1993.

[7] S.A. Nazarov, *Selfadjoint extensions of Dirichlet operators in spaces with functional weight*, Matematicheskiy sbornic 137(179), n. 2 (1988), 224–241 (in Russian).

[8] Yu.G. Shondin, *Quantum models in $\mathbb{R}^n$, connected with extension of the energy operator in the Pontrjagin's spaces*, Teoreticheskaya i matematicheskaya fizika, v. 74, n. 3 (1988), p. 331–344 (in Russian).

[9] Yu.G. Shondin, *Remark on the three-body problem with δ-interaction*, Teoreticheskaya i matematicheskaya fizika, v. 51, n. 2 (1982), p. 181–191 (in Russian).

[10] Yu.G. Shondin, *Generalized point-range interactions in $\mathbb{R}^3$ and connected with them models with the rational S-matrix 1, $l = 0$.* Teoreticheskaya i matematicheskaya fizika, v. 64, n. 3 (1985), p. 432–441 (in Russian).

[11] Yu.G. Shondin, *Generalized point-range interactions in $\mathbb{R}^3$ and connected with them models with the rational S-matrix 2, $l = 1$.* Teoreticheskaya i matematicheskaya fizika, v. 65, n. 1 (1985), p. 24–34 (in Russian).

[12] A. Dijksma, H. Langer, Yu. Shondin, and C. Zeinstra, *Self-adjoint operators with inner singularities in Pontryagin spaces*, Operator Theory and Related Topics, Vol. II (Odessa 1997), 105–175.English Translation: Operator Theory Adv. Appl., **118**, Birkhäuser, Basel 2000.

[13] A. Dijksma and H.S.V. de Snoo, *Symmetric and Self-adjoint Relations in Krein Spaces I*, Operator Theory Adv. Appl., **24**, Birkhäuser, Basel 1987, 145–166.

[14] B.S. Pavlov and I.Yu. Popov, *Acoustic model of zero width slits and hydrodynamical stability of boundary layer*, Teor Mat Fiz v. 86, n. 3, 1991 (in Russian).

[15] B.S. Pavlov and I.Yu. Popov, *Model difraction by infinitely narrow slit and extensions theory*, Vestink Leningr.Univ. Math, v. 16, 1984 (in Russian).

[16] B.S. Pavlov and I.Yu. Popov, *Scattering by responator with the small and pooint opennings*, Vestnik Lening. Univ., n. 13, 1984 (in Russian).

[17] B.S. Pavlov and I.Yu. Popov, *Running wave in the ring resonator*, Vestnik Lening. Univ., n. 4 1985 (in Russian).

[18] I.Yu. Popov, *A model of zero-width slits and the real diffraction problem*, Adv. and Applications, vol. 46, 1990, 195–196.

[19] I.Yu. Popov, *Construction of an inelastic scatterer in nanoelectronics by the extension-theory methods*, Adv. and Applications, vol. 46, 1990, 197–198.

[20] I.Yu. Popov, *Calculation of eigenfrequanceis of resonator connected through small opening with the use of the operator extensions theory*, Acoustic journal, v. 37, 1991, vup. 2, p. 380 (in Russian).

[21] Yu.A. Kuperin, K.A. Makarov, and B.S. Pavlov *The model of resonance scattering for complicated particles*, Teoreticheskaya i matematicheskaya fizika, v. 69, n. 1 (1986), p. 100–114 (in Russian).

[22] Yu. Karpeshina *Zero-range model of p-scattering*, Preprint ETH, Zürich, 1992, pp. 1–16.

[23] I. Andronov, *Zero-range Potential Model of a Protruding Stiffener*, J.Phys A, **32**, 1999, L231–238.

[24] P. Kurasov and K. Watanabe, *On rank one H_3 Perturbations of Positive Self-adjoint Operators*, in F. Gestesy, H. Holden, J. Jost, S. Peycha, M. Röckner, S. Scarlatti (Eds.), Stochastic Processes Physics and Geometry: New interplays II, CMS conference proceedings **29**. AMS, Providence, 2000, pp. 413–422.

[25] P. Kurasov, *H_{-n}-perturbations of Self-Adjoint Operators and Krein's Resolvent Formula*, Integr. Eq. Oper. Theory. **45** (2003), 437–460.

[26] P. Kurasov and K. Watanabe, *On H_{-n}-perturbations of Self-Adjoint Operators*, Partial Differential Equations and Spectral Theory (Clausthal, 2000), 179–196, Operator Theory and Appl., **126**, Birkhäuser, Basel, 2001.

[27] P. Kurasov, *Distribution theory for discontinuous test functions and differential operators with generalized coefficients*, J. Math. Anal. Appl., 201 (1996), 297–323.

[28] P. Kurasov, *Triplet extensions I: semibounded operators with defect one*, J. Anal. Math., **107** (2009), 251–286.

Yu.E. Karpeshina
Department of Mathematics
University of Alabama at Birmingham
1300 University blvd.
Birmingham
AL, 35294-1170, USA
e-mail: karpeshi@math.uab.edu

Operator Theory:
Advances and Applications, Vol. 214, 67–82
© 2011 Springer Basel AG

The Similarity Problem for Non-selfadjoint Operators with Absolutely Continuous Spectrum: Restrictions to Spectral Subspaces

Alexander V. Kiselev

Abstract. The similarity problem for restrictions of a non-selfadjoint operator with absolutely continuous spectrum to its spectral subspaces corresponding to arbitrary Borel subsets δ of the spectrum is considered, generalizing the results of [7, 11]. Necessary and sufficient conditions of such similarity are obtained in the form of a pair of integral estimates on $\delta \subset \mathbb{R}$. The results are then applied to the analysis of the one-dimensional non-selfadjoint Friedrichs model operator.

Mathematics Subject Classification (2000). Primary 47A10; Secondary 47A55.

Keywords. Non-selfadjoint operators, similarity problem, absolutely continuous spectrum.

1. Preliminaries

A non-selfadjoint operator L acting in the Hilbert space H is called similar to a self-adjoint operator A if there exists a bounded, boundedly invertible operator X in H, such that $L = X^{-1}AX$. The similarity problem thus arises, i.e., to ascertain whether a given non-selfadjoint operator L is similar to some self-adjoint operator A or not and to give preferably necessary and sufficient conditions of this.

A criterion for similarity of a general non-selfadjoint operator with real spectrum to a self-adjoint one was established in a form of a pair of integral estimates involving the resolvent of the operator in [15, 4] (see (3.2) below). Unfortunately, this criterion, although given in a very concise form, is hard to verify in applications. This is why it seems reasonable to rewrite it in an equivalent (and suitable for applications) form under some additional assumptions on the class of operators considered. One such assumption that has yielded rather interesting results in

---

The author gratefully acknowledges support received from the EPSRC grant EP/D00022X/2 and partial support from grants INTAS 05-1000008-7883 and RFBR 09-01-00515-a.

the case of additive non-selfadjoint perturbations of self-adjoint operators (see [7]) and in the case of non-selfadjoint extensions of symmetric operators with equal deficiency indices (see [9]) is that the spectrum of the operator L is absolutely continuous, i.e., the absolutely continuous spectral subspace of the operator coincides with the Hilbert space H (see Section 2 below for the definition and Remark 3.2). An essential ingredient of the approach suggested by us in [7] was the functional model of a non-selfadjoint operator.

Later, Malamud [11] has pointed out, that the result announced in [7] can be obtained in a different way, with no explicit use of functional model and related techniques.

In [9] it has been shown, that results similar to those of [7, 11] can be obtained in the case of non-selfadjoint extensions of symmetric operators with finite and equal deficiency indices. The technique used in this paper heavily involved the corresponding functional model, although its construction in this class of operators is quite different to the class considered in [7] and in the present paper. The existence of a concise model description of spectral projection to an arbitrary Borel set of absolutely continuous spectrum has further allowed in [9] to "localize" the similarity problem, i.e., to prove a criterion of similarity for the initial operator restricted to its spectral subspace, corresponding to an arbitrary Borel set of its a.c. spectrum. Results of this nature would seem to be impossible to derive using Malamud's technique.

The conditions of similarity to a selfadjoint operator derived in the present paper differ from ones that have been known previously (that is, the ones of [15, 4]) in that they are formulated in terms of objects defined on the real line only, whereas the existing ones were formulated in terms of half-planes of the complex plane. In a nutshell, using the functional model we have been able to pass to a limit in the Naboko-Van Casteren criterion under the only additional assumption that the operator L has absolutely continuous spectrum.

In the present paper, we present results analogous to [9] in the case of additive non-selfadjoint perturbations of self-adjoint operators, thus effectively generalizing the corresponding results of both [7] and [11].

The structure of the paper is as follows.

Since the functional model of a non-selfadjoint operator is of crucial importance for our approach and the proof of our main result relies heavily upon the symmetric form of the Nagy-Foiaş functional model due to Pavlov [20, 19] (see also the paper [14] by Naboko), we continue with a brief introduction to the main concepts and results obtained in this area in Section 2.

Section 3 contains our main result, which is a criterion of similarity of restrictions of non-selfadjoint operator under investigation to spectral subspaces corresponding to arbitrary Borel sets of the real line. Concise sufficient conditions follow almost immediately.

Finally, Section 4 demonstrates applicability of our results to analysis of the similarity problem for non-selfadjoint Friedrichs model operator on the real line.

2. The functional model

In the present section we briefly recall the functional model of a non-selfadjoint operator constructed in [16, 19] in the dissipative case and then extended in [12, 13, 14, 22] to the case of a wide class of non-dissipative operators. We consider a class of non-selfadjoint operators of the form [14] $L = A + iV$, where A is a self-adjoint operator in H defined on the domain $D(A)$ and the perturbation V admits the factorization $V = \alpha J \alpha/2$, where α is a non-negative self-adjoint operator in H and J is a unitary operator in an auxiliary Hilbert space E, defined as the closed range of the operator α: $E \equiv \overline{R(\alpha)}$. This factorization corresponds to the polar decomposition of the operator V. It can also be easily generalized[1] to the "node" case [25], where J acts in an auxiliary Hilbert space $\mathfrak{H}$ and $V = \alpha^* J\alpha/2$, α being an operator acting from H to $\mathfrak{H}$. In order that the expression $A + iV$ be meaningful, we impose the condition that V be (A)-bounded with relative bound less than 1, i.e., $D(A) \subset D(V)$ and for some a and b $(a < 1)$ the condition $\|Vu\| \le a\|Au\| + b\|u\|, \quad u \in D(A)$ is satisfied, see [6]. Then the operator L is well defined on the domain $D(L) = D(A)$.

Alongside with the operator L we are going to consider the maximal dissipative operator $L^{\|} = A + i\frac{\alpha^2}{2}$ and the one adjoint to it, $L^{-\|} \equiv L^{\|*} = A - i\frac{\alpha^2}{2}$. Since the functional model for the dissipative operator $L^{\|}$ will be used below, we require that $L^{\|}$ is completely non-selfadjoint, i.e., that it has no reducing self-adjoint parts. This requirement is not restrictive in our case due to Proposition 1 in [14].

We now briefly describe a construction of the self-adjoint dilation of the completely non-selfadjoint dissipative operator $L^{\|}$, following [16, 19], see also [14].

The characteristic function $S(\lambda)$ of the operator $L^{\|}$ is a contractive, analytic operator-valued function acting in the Hilbert space E, defined for $\operatorname{Im}\lambda > 0$ by

$$S(\lambda) = I + i\alpha(L^{-\|} - \lambda)^{-1}\alpha. \tag{2.1}$$

In the case of an unbounded α the characteristic function is first defined by the latter expression on the manifold $E \cap D(\alpha)$ and then extended by continuity to the whole space E. The definition given above makes it possible to consider $S(\lambda)$ for $\operatorname{Im}\lambda < 0$ with $S(\overline{\lambda}) = (S^*(\lambda))^{-1}$ provided that the inverse exists at the point λ. Finally, $S(\lambda)$ possesses boundary values on the real axis in the strong topology sense: $S(k) \equiv S(k + i0)$, $k \in \mathbb{R}$ (see [16]).

Consider the model space $\mathcal{H} = L_2\left(\begin{smallmatrix} I & S^* \\ S & I \end{smallmatrix}\right)$, which is defined in [19] (see also [17] for description of general coordinate-free models) as Hilbert space of two-component vector-functions $(\tilde g, g)$ on the axis $(\tilde g(k), g(k) \in E, k \in \mathbb{R})$ with metric

$$\left\langle \begin{pmatrix} \tilde g \\ g \end{pmatrix}, \begin{pmatrix} \tilde g \\ g \end{pmatrix} \right\rangle = \int_{-\infty}^{\infty} \left\langle \begin{pmatrix} I & S^*(k) \\ S(k) & I \end{pmatrix} \begin{pmatrix} \tilde g(k) \\ g(k) \end{pmatrix}, \begin{pmatrix} \tilde g(k) \\ g(k) \end{pmatrix} \right\rangle_{E\oplus E} dk.$$

It is assumed here that the set of two-component functions has been factored by the set of elements with norm equal to zero. Although we consider $(\tilde g, g)$ as a

[1] Results of Section 3 below admit natural generalization to the "node" case. We have chosen to refrain from including the corresponding details in order to simplify the reading.

symbol only, the formal expressions $g_- := (\tilde{g} + S^*g)$ and $g_+ := (S\tilde{g} + g)$ (the motivation for the choice of notation is self-evident from what follows) can be shown to represent some true $L_2(E)$-functions on the real line. In what follows we plan to deal mostly with these functions.

Define the following orthogonal subspaces in $\mathcal{H}$:

$$D_- \equiv \begin{pmatrix} 0 \\ H^2_-(E) \end{pmatrix}, \; D_+ \equiv \begin{pmatrix} H^2_+(E) \\ 0 \end{pmatrix}, \; K \equiv \mathcal{H} \ominus (D_- \oplus D_+),$$

where $H^2_{+(-)}(E)$ denotes the Hardy class [16] of analytic functions f in the upper (lower) half-plane taking values in the Hilbert space E. These subspaces are "incoming" and "outgoing" subspaces, respectively, in the sense of [10].

The subspace K can be described as $K = \{(\tilde{g}, g) \in \mathcal{H} : \; g_- \equiv \tilde{g} + S^*g \in H^2_-(E), g_+ \equiv S\tilde{g} + g \in H^2_+(E)\}$. Let P_K be the orthogonal projection of the space $\mathcal{H}$ onto K, then

$$P_K \begin{pmatrix} \tilde{g} \\ g \end{pmatrix} = \begin{pmatrix} \tilde{g} - P_+(\tilde{g} + S^*g) \\ g - P_-(S\tilde{g} + g) \end{pmatrix},$$

where $P_\pm$ are the orthogonal Riesz projections of the space $L_2(E)$ onto $H^2_\pm(E)$.

The following Theorem holds [16, 19]:

Theorem 2.1. *The operator $(L^{\|} - \lambda_0)^{-1}$ is unitarily equivalent to the operator $P_K(k - \lambda_0)^{-1}|_K$ in the space K for all $\lambda_0, \operatorname{Im} \lambda_0 < 0$.*

This means, that the operator of multiplication by k in $\mathcal{H}$ serves as a minimal $(\operatorname{clos}_{\operatorname{Im}\lambda\neq 0}(k - \lambda)^{-1}K = \mathcal{H})$ self-adjoint dilation [16] of the operator $L^{\|}$.

Provided that the non-real spectrum of the operator L is countable, the characteristic function of the operator L is defined for $\operatorname{Im} \lambda \neq 0$ by the expression $\Theta(\lambda) \equiv I + iJ\alpha(L^* - \lambda)^{-1}\alpha$ and under an additional assumption that V is a relatively compact perturbation[2] can be shown to be a meromorphic, J-contractive $(\Theta^*(\lambda)J\Theta(\lambda) \leq J, \quad \operatorname{Im} \lambda > 0)$ operator-function [3]. The characteristic function $\Theta(\lambda)$ admits [1, 12] a factorization (also called Ginzburg-Potapov factorization of a J-contractive function [2]) in the form of a ratio of two bounded analytic operator-functions (in the corresponding half-planes $\operatorname{Im} \lambda < 0, \; \operatorname{Im} \lambda > 0$) triangular with respect to decomposition of the space E into the orthogonal sum $E = (\mathcal{X}_+E) \oplus (\mathcal{X}_-E)$, $\mathcal{X}_\pm := (I \pm J)/2$:

$$\begin{aligned} \Theta(\lambda) &= \Theta_1'^*(\overline{\lambda})(\Theta_2'^*)^{-1}(\overline{\lambda}), \; \operatorname{Im} \; \lambda > 0; \\ \Theta(\lambda) &= \Theta_2^*(\overline{\lambda})(\Theta_1^*)^{-1}(\overline{\lambda}), \; \operatorname{Im} \; \lambda < 0, \end{aligned} \tag{2.2}$$

where the factors $\Theta_{1,2}$ and $\Theta'_{1,2}$ are introduced as follows [13]:

$$\begin{aligned} \Theta_1(\lambda) &= \mathcal{X}_- + S(\lambda)\mathcal{X}_+, \qquad & \Theta_2(\lambda) &= \mathcal{X}_+ + S(\lambda)\mathcal{X}_-; \\ \Theta_1'(\lambda) &= \mathcal{X}_- + S^*(\overline{\lambda})\mathcal{X}_+, \qquad & \Theta_2'(\lambda) &= \mathcal{X}_+ + S^*(\overline{\lambda})\mathcal{X}_-, \end{aligned} \tag{2.3}$$

and $S(\lambda)$ is the characteristic function of the dissipative operator $L^{\|}$.

[2]This assumption guarantees that the non-real spectrum of L is discrete.

Following [13], we define the linear sets $\hat{N}_\pm$ in $\mathcal{H}$ as follows:

$$\hat{N}_\pm \equiv \left\{ \begin{pmatrix} \tilde{g} \\ g \end{pmatrix} : \begin{pmatrix} \tilde{g} \\ g \end{pmatrix} \in \mathcal{H}, P_\pm \left(\Theta_1'^* \tilde{g} + \Theta_2^* g \right) \equiv P_\pm \left(\mathcal{X}_+ g_+ + \mathcal{X}_- g_- \right) = 0 \right\} \tag{2.4}$$

and introduce subspaces $N_\pm = \operatorname{clos} P_K \hat{N}_\pm$. Then, as it is shown in [14], one gets for Im $\lambda < 0$ (Im $\lambda > 0$) and $(\tilde{g}, g) \in \hat{N}_{-(+)}$, respectively:

$$(L - \lambda)^{-1} P_K \begin{pmatrix} \tilde{g} \\ g \end{pmatrix} = P_K \frac{1}{k - \lambda} \begin{pmatrix} \tilde{g} \\ g \end{pmatrix}. \tag{2.5}$$

Conversely, the property (2.5) for Im $\lambda < 0$ (Im $\lambda > 0$) guarantees that the vector $(\tilde{g}, g)$ belongs to the set $\hat{N}_{-(+)}$.

Absolutely continuous and singular subspaces of the non-selfadjoint operator L were defined in [12]: let $N \equiv \hat{N}_+ \cap \hat{N}_-$, $\tilde{N}_\pm \equiv P_K \hat{N}_\pm$, $\tilde{N}_e \equiv \tilde{N}_+ \cap \tilde{N}_-$. Then[3]

$$N_e \equiv \operatorname{clos} \left(\tilde{N}_+ \cap \tilde{N}_- \right) = \operatorname{clos} P_K N \equiv \operatorname{clos} \tilde{N}_e; \quad N_i \equiv K \ominus N_e(L^*), \tag{2.6}$$

where $N_e(L^*)$ denotes the absolutely continuous subspace of the operator L^*, which can be easily described in similar way in terms of the same model space $\mathcal{H}$.

One can also ascertain that the linear sets $\tilde{N}_\pm$ can be characterized in terms, independent of the functional model, in the following way:

$$\tilde{N}_{+(-)} = \left\{ u \in H : \mathcal{X}_{+(-)} \alpha (L - \lambda)^{-1} u \in H^2_{+(-)}(E) \right\}. \tag{2.7}$$

Here $\mathcal{X}_{+(-)} \alpha (L - \lambda)^{-1} u$[4] is treated as an analytic vector function of $\lambda \in \mathbb{C}_{+(-)}$ taking values in the auxiliary Hilbert space E. It can be verified [14] that the projections $\mathcal{X}_\pm$ can be dropped altogether in the definition (2.7). The existence of this description gives ground to calling the vectors belonging to the named linear sets "smooth".

The definition (2.6) in the case of maximal dissipative operators leads to the same subspace as the classical definition by L.A. Sahnovich [23] (the latter definition introduces the absolutely continuous subspace as the maximal invariant subspace reducing the operator L to an operator with purely outer characteristic function) and was later developed by V.A. Ryzhov (in the case of more general non-dissipative operators) [22] and A.S. Tikhonov [24] (the so-called weak definition of the absolutely continuous subspace). Recently it turned out that in the dissipative situation the weak definition coincides with the strong one (2.6) (see [21]).

The subspaces $N_\pm(L^*)$ for the adjoint to L operator L^* are defined in a similar way using the same model representation.

---

[3] The linear set $\tilde{N}_e$ is called the set of "smooth" vectors of the operator L (see [14]).

[4] That is, analytic continuations of the vector $\mathcal{X}_{+(-)} \alpha (L - \lambda)^{-1} u$ from the domain of analyticity of the resolvent to the half-plane $\mathbb{C}_{+(-)}$.

3. Similarity problem for additive non-selfadjoint perturbations with absolutely continuous spectrum

Definition 3.1. We call a nonselfadjoint operator A acting in Hilbert space H *an operator with absolutely continuous spectrum* if its absolutely continuous subspace $N_e(A)$ coincides with H.

Remark 3.2. It has to be noted that the requirement $N_e(A) = H$ doesn't actually guarantee that the spectrum of the operator A is *purely* absolutely continuous. Due to the possibility that the absolutely continuous and singular spectral subspaces may intersect, one might face the following situation: $N_e(A) = H$; $N_i(A) \neq \{0\}$ and $N_i(A) \subset N_e(A)$. See [26] for one rather transparent example of this case. Nevertheless, due to the fact that $N_i(A^*) \equiv H \ominus N_e(A)$, one easily sees that in the situation of $N_e(A) = H$ the singular spectral subspace of the *adjoint* operator A^* is trivial.

The spectral projection $\mathcal{P}_\delta$ corresponding to the "portion" of the absolutely continuous spectrum contained in a Borel set δ was constructed in model terms in [13]. Namely, the following result holds:

Proposition 3.3. *Suppose that L is a completely non-selfadjoint operator with absolutely continuous spectrum. For any Borel set $\delta \subset \mathbb{R}$ put*

$$\mathcal{P}_\delta P_K \begin{pmatrix} \tilde{g} \\ g \end{pmatrix} = P_K \mathcal{X}_\delta \begin{pmatrix} \tilde{g} \\ g \end{pmatrix}, \tag{3.1}$$

where $\binom{\tilde{g}}{g} \in N$ and $\mathcal{X}_\delta$ is the operator of componentwise multiplication by the characteristic function of the set δ. For the operator $\mathcal{P}_\delta$ defined by (3.1) on the set of smooth vectors $\tilde{N}_e$ the following assertions hold:

(i) $\mathcal{P}_\delta \tilde{N}_e \subset \tilde{N}_e$;
(ii) $(L-\lambda_0)^{-1}\mathcal{P}_\delta = \mathcal{P}_\delta (L-\lambda_0)^{-1}, \quad \operatorname{Im}\lambda_0 \neq 0$;
(iii) $\mathcal{P}_\delta \mathcal{P}_{\delta'} = \mathcal{P}_{\delta\cap\delta'}, \quad \delta, \delta' \subset \mathbb{R}$;
(iv) $\mathcal{P}_\delta u \longrightarrow u$ *as* $\delta \to (-\infty,\infty), \quad u \in \tilde{N}_e$ *(in fact, as* $1-\mathcal{X}_\delta \to 0$ *in* $L_\infty(\mathbb{R})$*)*;
(v) $\mathcal{P}_\delta u = \lim_{\varepsilon\to+0} \frac{1}{2\pi i}\int_\delta [(L-k-i\varepsilon)^{-1} - (L-k+i\varepsilon)^{-1}] u\,dk, \quad u \in \tilde{N}_e$.

We remark that the assertion (v) above establishes the connection between the definition of a spectral projection in terms of the functional model with the usual approach to the definition of spectral projections based on the Riesz integral for the resolvent, and thus the term "spectral projection" is justified.

Based on this result, in the present Section we obtain conditions, necessary and sufficient for the restrictions of the non-selfadjoint operator A of the class considered in this paper to its spectral subspaces (i.e., to subspaces of the form clos $\mathcal{P}_\delta N_e$) to be similar to selfadjoint operators. We assume throughout that the operator L is an operator with absolutely continuous spectrum.

It's proved in [15, 4], that a non-selfadjoint operator A acting in Hilbert space H, the spectrum $\sigma(A)$ of which is a subset of real axis, is similar to a selfadjoint operator if and only if there exists a finite constant C such that on any vector $u \in H$ the following estimates hold:

$$\begin{aligned} &\sup_{\varepsilon>0} \varepsilon \int_{-\infty}^{\infty} \|(L-k-i\varepsilon)^{-1}u\|^2 dk \le C\|u\|^2 \\ &\sup_{\varepsilon>0} \varepsilon \int_{-\infty}^{\infty} \|(L^*-k-i\varepsilon)^{-1}u\|^2 dk \le C\|u\|^2 \\ &\sup_{\varepsilon>0} \varepsilon \int_{-\infty}^{\infty} \|(L-k+i\varepsilon)^{-1}u\|^2 dk \le C\|u\|^2 \\ &\sup_{\varepsilon>0} \varepsilon \int_{-\infty}^{\infty} \|(L^*-k+i\varepsilon)^{-1}u\|^2 dk \le C\|u\|^2 \end{aligned} \tag{3.2}$$

Furthermore, it's not hard to see that the first pair of estimates above is equivalent to the second pair (i.e., the first pair of estimates in (3.2) holds on any $u \in H$ iff the second pair of estimates holds on any $u \in H$). This makes it possible to prove the following

Theorem 3.4. *Provided that the spectrum of L is absolutely continuous, the following assertions are equivalent*[5]*:*

(a) *The restriction of the operator L to its spectral invariant subspace, corresponding to a Borel set $\delta \subset \mathbb{R}$, is similar to a self-adjoint operator acting in* clos $\mathcal{P}_\delta N_e$*;*

(b) *There exists a $C < \infty$ such that for any $u \in$ clos $\mathcal{P}_\delta N_e$ the following estimates hold:*

$$\int_\delta \Big((\Theta(k-i0)J\Theta^*(k-i0) - J)\mathcal{X}_+\alpha(L^{-\|}-k-i0)^{-1}u, \\ \mathcal{X}_+\alpha(L^{-\|}-k-i0)^{-1}u \Big) dk \le C\|u\|^2$$

$$\int_\delta \Big((J - \Theta^*(k+i0)J\Theta(k+i0))\mathcal{X}_-\alpha(L^{-\|}-k-i0)^{-1}u, \\ \mathcal{X}_-\alpha(L^{-\|}-k-i0)^{-1}u \Big) dk \le C\|u\|^2;$$

(c) *There exists a $C < \infty$ such that for any $u \in$ clos $\mathcal{P}_\delta N_e$ the following estimates hold:*

$$\int_\delta \Big((J - \Theta(k+i0)J\Theta^*(k+i0))\mathcal{X}_-\alpha(L^{\|}-k+i0)^{-1}u, \\ \mathcal{X}_-\alpha(L^{\|}-k+i0)^{-1}u \Big) dk \le C\|u\|^2$$

$$\int_\delta \Big((\Theta^*(k-i0)J\Theta(k-i0) - J)\mathcal{X}_+\alpha(L^{\|}-k+i0)^{-1}u, \\ \mathcal{X}_+\alpha(L^{\|}-k+i0)^{-1}u \Big) dk \le C\|u\|^2.$$

---

[5]Under our assumptions the characteristic function $\Theta(\lambda)$ might have no boundary values on the real line on its own; however, the boundary values of its J-forms $J - \Theta^* J\Theta$ and $\Theta J\Theta^* - J$ do exist due to [14]. The corresponding boundary values below should be understood accordingly.

Proof. Our first goal is to rewrite estimates (3.2) in terms of the functional model for additive non-selfadjoint perturbations of self-adjoint operators described in Section 2 above. We will see that provided that the spectrum of the operator in question is absolutely continuous it's possible to compute the limits of the left-hand sides in the resulting estimates when $\varepsilon \to 0$ precisely, thus making it possible to replace (3.2) with much more simple conditions, formulated in terms of objects defined on the real line only.

This passage from the uniform integral estimates in the half-plane to the integral estimates on the boundary values simplifies matters in the case of one-dimensional Friedrichs model (see [18, 8], cf. [15]). As shown in [8], our approach reduces the similarity problem to the question of boundedness of certain singular integral operators. The latter could be examined by standard methods of analysis yielding necessary and sufficient conditions for the similarity of the operator of the Friedrichs model to a self-adjoint one. It is also worth mentioning that in the process of this limit procedure we reduce the consideration to analytic functions taking values in Hilbert spaces of (potentially) lower dimensions.

The following proposition holds:

Proposition 3.5. *The estimates in* (3.2), *considered on the vectors* $u \in \operatorname{clos} \mathcal{P}_\delta N_e$, *are one-to-one equivalent to the following ones:*

$$\begin{aligned}
\left\| P_+ \begin{pmatrix} \tilde g + S^* g \\ -(S\tilde g + g) \end{pmatrix} \right\|^2_{\mathcal{H}} &\le C \left\| P_K \begin{pmatrix} \tilde g \\ g \end{pmatrix} \right\|^2_{\mathcal{H}} \\
\left\| \begin{pmatrix} P_+(\tilde g + S^* g) - c_2(k) \\ -P_+(S\tilde g + g) + S(k) c_2(k) \end{pmatrix} \right\|^2_{\mathcal{H}} &\le C \left\| P_K \begin{pmatrix} \tilde g \\ g \end{pmatrix} \right\|^2_{\mathcal{H}} \\
\left\| P_- \begin{pmatrix} \tilde g + S^* g \\ -(S\tilde g + g) \end{pmatrix} \right\|^2_{\mathcal{H}} &\le C \left\| P_K \begin{pmatrix} \tilde g \\ g \end{pmatrix} \right\|^2_{\mathcal{H}} \\
\left\| \begin{pmatrix} -P_-(\tilde g + S^* g) + S^*(k) c_1(k) \\ P_-(S\tilde g + g) - c_1(k) \end{pmatrix} \right\|^2_{\mathcal{H}} &\le C \left\| P_K \begin{pmatrix} \tilde g \\ g \end{pmatrix} \right\|^2_{\mathcal{H}},
\end{aligned} \tag{3.3}$$

where $\binom{\tilde g}{g} \in \mathcal{X}_\delta N$ *and*

$$\begin{aligned}
T_1(\lambda) &\equiv [\mathcal{X}_- + S^*(\overline{\lambda})\mathcal{X}_+]^{-1} \\
T_2(\lambda) &\equiv [\mathcal{X}_+ + \mathcal{X}_- S(\lambda)]^{-1} \\
c_1(\lambda) &\equiv T_1(\lambda)(P_-(\tilde g + S^* g)(\lambda) + P_-(S\tilde g + g)(\lambda)) \\
c_2(\lambda) &\equiv T_2(\lambda)(P_+(\tilde g + S^* g)(\lambda) + P_+(S\tilde g + g)(\lambda)),
\end{aligned}$$

the boundary values of $c_1(\lambda)$ *and* $c_2(\lambda)$ *almost everywhere on the real line existing for all* $(\tilde g, g) \in \mathcal{X}_\delta N$.

The proof does not differ from the case of non-selfadjoint extensions of symmetric operators and therefore we refer the reader to the paper [9].

In order to complete the proof of Theorem 3.4, we need to rewrite the estimates obtained by virtue of Proposition 3.5 in terms of the initial Hilbert space

H and of the operators in it. To this end, we will first rewrite our estimates in terms of $\mathfrak{H}$, the three-component unitary image of $\mathcal{H}$, see [19, 14]. The space $\mathfrak{H} \equiv D_- \oplus H \oplus D_+$ consists of three-component vector-functions $(\tilde{v}_-, u, \tilde{v}_+)$, where $\tilde{v}_- \in L_2(\mathbb{R}_-; E), \tilde{v}_+ \in L_2(\mathbb{R}_+; E)$ and $u \in H$. The unitary operator (see [14]) that maps $\mathfrak{H}$ onto $\mathcal{H}$ is given by the following formulae:

$$\tilde{g} + S^* g = -\frac{1}{\sqrt{2\pi}} \alpha(L^{\|} - k + i0)^{-1} u + S^*(k) v_-(k) + v_+(k)$$
$$S\tilde{g} + g = -\frac{1}{\sqrt{2\pi}} \alpha(L^{-\|} - k - i0)^{-1} + v_-(k) + S(k) v_+(k),$$

where[6]

$$v_\pm(k) \equiv \frac{1}{\sqrt{2\pi}} \int e^{ik\xi} \tilde{v}_\pm(\xi) d\xi \in H_2^\pm(E)$$

by the Paley–Wiener theorem [5].

We are going to use this mapping extensively. First of all, note that the fact that $\binom{\tilde{g}}{g} \in N$ in the model representation is equivalent to

$$\begin{aligned} \mathcal{X}_-(\tilde{g} + S^* g) &= 0 \\ \mathcal{X}_+(S\tilde{g} + g) &= 0 \end{aligned}, \tag{3.4}$$

which is of course also true for the subspace we are considering, $\mathcal{P}_\delta N$. Next, for the latter subspace we clearly have

$$\begin{aligned} \mathcal{X}_\delta(\tilde{g} + S^* g) &= \tilde{g} + S^* g \\ \mathcal{X}_\delta(S\tilde{g} + g) &= S\tilde{g} + g \end{aligned}, \tag{3.5}$$

and finally,

$$\begin{aligned} \mathcal{X}_{+(-)} \alpha(L^{\|(-\|)} + (-)i0)^{-1} u &\in H^2_{-(+)}(E) \\ \mathcal{X}_+ v_-(k) &= 0 \quad \text{for a.a. } k \\ \mathcal{X}_- v_+(k) &= 0 \quad \text{for a.a. } k \\ \mathcal{X}_- S^*(k) v_-(k) &= \frac{1}{\sqrt{2\pi}} \mathcal{X}_- \alpha(L^{\|} - k + i0)^{-1} u \quad \text{for a.a. } k \\ \mathcal{X}_+ S(k) v_+(k) &= \frac{1}{\sqrt{2\pi}} \mathcal{X}_+ \alpha(L^{-\|} - k - i0)^{-1} u \quad \text{for a.a. } k, \end{aligned} \tag{3.6}$$

where we have used (3.4) and the orthogonality of $H^2_+(E)$ and $H^2_-(E)$ in $L^2(E)$.

We prove now that the assertions (a) and (c) of Theorem 3.4 are equivalent. In order to do so we need to show, that the third and fourth estimates in the statement of Proposition 3.5 are respectively equivalent to the ones provided by the assertion (c) of Theorem 3.4.

---

[6] We assume here that the functions $\tilde{v}_-, \tilde{v}_+$ have been extended by zero to the complementary semiaxes.

We begin with the third estimate. One immediately obtains:

$$\left\| \begin{pmatrix} P_-(\tilde g + S^* g) \\ -P_-(S\tilde g + g) \end{pmatrix} \right\|^2_{\mathcal{H}}$$
$$= \int_\delta dk \Big(\|P_-(\tilde g + S^* g)\|^2 + \|P_-(S\tilde g + g)\|^2 - 2\,\mathrm{Re}\big(SP_-(\tilde g + S^* g), P_-(S\tilde g + g)\big)\Big),$$

where we have used (3.4), (3.5) and the following simple observation:

$$\int_{\mathbb{R}} (P_- f_1(k), f_2(k))dk = \int_{\mathbb{R}} (P_- f_1(k), P_- f_2(k))dk$$
$$= \int_\delta (f_1(k), P_- f_2(k))dk = \int_\delta (P_- f_1(k), f_2(k))dk \quad \text{when } f_1(k) = \mathcal{X}_\delta f_1(k).$$

Direct computation now shows that the third estimate of Proposition 3.5 is equivalent to the following one:

$$\int_\delta dk \left(\|\mathcal{X}_- v_-\|^2 - \|\mathcal{X}_+ S^* v_-\|^2 \right) \tag{3.7}$$
$$+ 2\,\mathrm{Re} \int_\delta dk \left(\frac{1}{\sqrt{2\pi}} \mathcal{X}_+ \alpha (L^{\|} - k + i0)^{-1} u, (S^*(k)\mathcal{X}_- - \mathcal{X}_+ S^*(k))\, v_-(k) \right)$$
$$\le C\|u\|^2,$$

where we have taken into account that [14]

$$\int_\delta \left\| \frac{1}{\sqrt{2\pi}} \mathcal{X}_+ \alpha (L^{\|} - k + i0)^{-1} u \right\|^2 dk \le C\|u\|^2.$$

The conditions (3.6) when applied to (3.7) show the equivalence of the latter estimate to

$$\int_\delta dk \Big((\mathcal{X}_- S \mathcal{X}_-)^{-1} (I - SS^*) (\mathcal{X}_- S^* \mathcal{X}_-)^{-1} \mathcal{X}_- \alpha (L^{\|} - k + i0)^{-1} u\ ,$$
$$\mathcal{X}_- \alpha (L^{\|} - k + i0)^{-1} u \Big) \le C\|u\|^2,$$

where $\mathcal{X}_- S \mathcal{X}_-$ is treated as a bounded linear operator on $\mathcal{X}_- E$ for a.a. k. On the basis of Hilbert identity it can be now shown, that

$$(\mathcal{X}_- S(\lambda) \mathcal{X}_-)^{-1} = \mathcal{X}_- \Theta(\lambda) \mathcal{X}_-.$$

Then

$$\Big((\mathcal{X}_- S(\lambda)\mathcal{X}_-)^{-1} (I - S(\lambda) S^*(\lambda)) \,(\mathcal{X}_- S^*(\lambda) \mathcal{X}_-)^{-1} \mathcal{X}_- \alpha (L^{\|} - \overline{\lambda})^{-1} u,$$
$$\mathcal{X}_- \alpha (L^{\|} - \overline{\lambda})^{-1} u \Big)$$
$$= \Big(\mathcal{X}_- \Theta(\lambda) \mathcal{X}_- \Theta_2(\lambda) J (J - \Theta(\lambda) J \Theta^*(\lambda))\, J \Theta_2^*(\lambda) \mathcal{X}_- \Theta^*(\lambda) \mathcal{X}_- \alpha (L^{\|} - \overline{\lambda})^{-1} u,$$
$$\mathcal{X}_- \alpha (L^{\|} - \overline{\lambda})^{-1} u \Big)$$
$$= \Big(J (J - \Theta(\lambda) J \Theta^*(\lambda))\, J \mathcal{X}_- \alpha (L^{\|} - \overline{\lambda})^{-1} u, \mathcal{X}_- \alpha (L^{\|} - \overline{\lambda})^{-1} u \Big),$$

since, clearly, $\Theta_2^*(\lambda)\mathcal{X}_-\Theta^*(\lambda)\mathcal{X}_- = \mathcal{X}_-$. The last identity leads to the desired estimate.

Analogous computations based on (3.4), (3.5) and (3.6) applied to the fourth estimate of the Lemma 3.5 show that the latter is equivalent to the following one:

$$\int_\delta dk((\mathcal{X}_- + S(k)\mathcal{X}_+)^{-1}(I - SS^*)(\mathcal{X}_- + \mathcal{X}_+S^*(k))^{-1}\mathcal{X}_+\alpha(L^{\|} - k + i0)^{-1}u,\\ \mathcal{X}_+\alpha(L^{\|} - k + i0)^{-1}u) \le C\|u\|^2. \quad (3.8)$$

Taking into account that

$$\mathcal{X}_- + \mathcal{X}_+S^*(\lambda) = (\mathcal{X}_- + S(\lambda)\mathcal{X}_+)^* = \Theta_1^*(\lambda),$$

we have:

$$\begin{aligned}
&((\mathcal{X}_- + S(\lambda)\mathcal{X}_+)^{-1}(I - S(\lambda)S^*(\lambda))(\mathcal{X}_- + \mathcal{X}_+S^*(\lambda))^{-1}\mathcal{X}_+\alpha(L^{\|} - \overline{\lambda})^{-1}u,\\
&\qquad \mathcal{X}_+\alpha(L^{\|} - \overline{\lambda})^{-1}u)\\
&= (\Theta_1^{-1}(\lambda)\Theta_2(\lambda)J(J - \Theta(\lambda)J\Theta^*(\lambda))J\Theta_2^*(\lambda)\Theta_1^{*-1}(\lambda)\mathcal{X}_+\alpha(L^{\|} - \overline{\lambda})^{-1}u,\\
&\qquad \mathcal{X}_+\alpha(L^{\|} - \overline{\lambda})^{-1}u)\\
&= (\Theta^*(\overline{\lambda})J(J - \Theta(\lambda)J\Theta^*(\lambda))J\Theta(\overline{\lambda})\mathcal{X}_+\alpha(L^{\|} - \overline{\lambda})^{-1}u, \mathcal{X}_+\alpha(L^{\|} - \overline{\lambda})^{-1}u)\\
&= ((\Theta^*(\overline{\lambda})J\Theta(\overline{\lambda}) - J)\mathcal{X}_+\alpha(L^{\|} - \overline{\lambda})^{-1}u, \mathcal{X}_+\alpha(L^{\|} - \overline{\lambda})^{-1}u),
\end{aligned}$$

which finishes the proof of the equivalence of the assertions (a) and (c) of the Theorem 3.4.

The equivalence of assertions (a) and (b) is shown in analogous fashion, so we omit the corresponding calculations here. □

The results obtained look still quite complicated since the integral estimates of Theorem 3.4 involve the boundary values of the resolvent of the dissipative operator $L^{\|}$ and its adjoint, rather then the boundary values of the operators L and L^* like the conditions (3.2). Therefore we now prove a modification of the Theorem 3.4. Namely, the following result holds:

Theorem 3.6. *Provided that the spectrum of L is absolutely continuous, the following assertions are equivalent:*

(a) *The restriction of the operator L to its spectral invariant subspace, corresponding to a Borel set $\delta \subset \mathbb{R}$, is similar to a self-adjoint operator acting in* clos $\mathcal{P}_\delta N_e$*;*

(b) *For any $u \in$ clos $\mathcal{P}_\delta N_e$ the following estimates hold:*

$$\int_\delta ((I - S^*(k)S(k))\mathcal{X}_+\alpha(L - k - i0)^{-1}u, \mathcal{X}_+\alpha(L - k - i0)^{-1}u)dk \le C\|u\|^2$$

$$\int_\delta ((I - S^*(k)S(k))\mathcal{X}_-\alpha(L^* - k - i0)^{-1}u, \mathcal{X}_-\alpha(L^* - k - i0)^{-1}u)dk \le C\|u\|^2;$$

(c) *For any* $u \in \operatorname{clos} \mathcal{P}_\delta N_e$ *the following estimates hold:*

$$\int_\delta ((I - S(k)S^*(k))\mathcal{X}_-\alpha(L - k + i0)^{-1}u, \mathcal{X}_-\alpha(L - k + i0)^{-1}u)dk \leq C\|u\|^2$$

$$\int_\delta ((I - S(k)S^*(k))\mathcal{X}_+\alpha(L^* - k + i0)^{-1}u, \mathcal{X}_+\alpha(L^* - k + i0)^{-1}u)dk \leq C\|u\|^2.$$

Proof. This theorem is proved by direct computation. For example, for the first estimate of the assertion (b) of Theorem 3.4 one has:

$$\begin{aligned}((\Theta(\overline{\lambda})J\Theta^*(\overline{\lambda}) - J)\mathcal{X}_+\alpha(L^{-\|} - \lambda)^{-1}u, \mathcal{X}_+\alpha(L^{-\|} - \lambda)^{-1}u)\\ = (\Theta_1^*(\overline{\lambda})\mathcal{X}_+(\Theta(\overline{\lambda})J\Theta^*(\overline{\lambda}) - J)\mathcal{X}_+\Theta_1(\lambda)\alpha(L - \lambda)^{-1}u, \alpha(L - \lambda)^{-1}u),\end{aligned}$$

since

$$\alpha(L^{-\|} - \lambda)^{-1} = \Theta_1(\lambda)\alpha(L - \lambda)^{-1}.$$

Then

$$\begin{aligned}&\Theta_1^*(\overline{\lambda})\mathcal{X}_+(\Theta(\overline{\lambda})J\Theta^*(\overline{\lambda}) - J)\mathcal{X}_+\Theta_1(\lambda)\\ &\quad= (\mathcal{X}_+ - \mathcal{X}_+S^*(\lambda)\mathcal{X}_-)J(\mathcal{X}_+ - \mathcal{X}_-S(\lambda)\mathcal{X}_+) - \Theta_1^*(\lambda)\mathcal{X}_+J\mathcal{X}_+\Theta_1(\lambda)\\ &\quad= (\mathcal{X}_+ - \mathcal{X}_+S^*(\lambda)\mathcal{X}_-)(\mathcal{X}_+ + \mathcal{X}_-S(\lambda)\mathcal{X}_+)\\ &\qquad - (\mathcal{X}_- + \mathcal{X}_+S^*(\lambda))\mathcal{X}_+(\mathcal{X}_- + S(\lambda)\mathcal{X}_+)\\ &\quad= \mathcal{X}_+ - \mathcal{X}_+S^*(\lambda)\mathcal{X}_-S(\lambda)\mathcal{X}_+ - \mathcal{X}_+S^*(\lambda)\mathcal{X}_+S(\lambda)\mathcal{X}_+\\ &\quad= \mathcal{X}_+(I - S^*(\lambda)S(\lambda))\mathcal{X}_+,\end{aligned}$$

where we have used the fact that by the Hilbert identity

$$\Theta^*(\overline{\lambda})\mathcal{X}_+\Theta_1(\lambda) = \mathcal{X}_+ - \mathcal{X}_-S(\lambda)\mathcal{X}_+.$$

In the case of the other respective pairs of estimates the proof is carried out similarly. □

Corollary 3.7. *Provided that the spectrum of the operator* L *is absolutely continuous, either of the following conditions is sufficient for the restriction of the operator* L *to its spectral invariant subspace, corresponding to a Borel set* $\delta \subset \mathbb{R}$, *to be similar to a self-adjoint operator acting in* $\operatorname{clos} \mathcal{P}_\delta N_e$*:*

(a) *There exists a constant* $C < \infty$ *such that for all* $u \in \operatorname{clos} \mathcal{P}_\delta N_e$ *the following estimates hold:*

$$\begin{aligned}\int_\delta \left\|\mathcal{X}_+\alpha(L - k - i0)^{-1}u\right\|^2 dk \leq C\|u\|^2\\ \int_\delta \left\|\mathcal{X}_-\alpha(L^* - k - i0)^{-1}u\right\|^2 dk \leq C\|u\|^2\end{aligned} \qquad (3.9)$$

(b) *There exists a constant* $C < \infty$ *such that for all* $u \in \operatorname{clos} \mathcal{P}_\delta N_e$ *the following estimates hold:*

$$\begin{aligned} \int_\delta \left\| \mathcal{X}_- \alpha (L - k + i0)^{-1} u \right\|^2 dk \le C \|u\|^2 \\ \int_\delta \left\| \mathcal{X}_+ \alpha (L^* - k + i0)^{-1} u \right\|^2 dk \le C \|u\|^2. \end{aligned} \tag{3.10}$$

In the last section of the present paper we are going to apply the results obtained above (more specifically, the result of Corollary 3.7) to the analysis of the similarity problem for the operator of one-dimensional non-selfadjoint Friedrichs model.

4. Application: Friedrichs model operator

We consider the operator acting on its natural domain in the Hilbert space $L_2(\mathbb{R})$ and defined by the formula

$$(Lu)(x) = xu(x) + \psi(x) \int u(t)\overline{\varphi(t)}\, dt, \qquad \varphi, \psi \in L_2(\mathbb{R}). \tag{4.1}$$

The determinant of perturbation $D(\lambda)$ in this case is given by the following expression: $D(\lambda) = 1 + \int \overline{\varphi(t)}\psi(t)(t-\lambda)^{-1}\, dt$. In order to simplify the calculation of the operators α and $\mathcal{X}_\pm$ we restrict ourselves to the case of orthogonal functions φ, ψ: $(\varphi, \psi) = 0$. The corresponding calculations can clearly be carried out in the general case as well.

Denote the class of the functions f analytic in the upper (lower) half-plane and satisfying the condition

$$\sup_{\varepsilon>0\,(\varepsilon<0)} \int |f(k+i\varepsilon)|^p \frac{dk}{1+k^2} < \infty$$

by $H_+^{p,\mathrm{loc}}$ $(H_-^{p,\mathrm{loc}})$.

The following Lemma, characterizing the structure of the spectrum of the operator under investigation, can be derived from (2.7) by a rather straightforward calculation:

Lemma 4.1.

(i) *Let the spectrum of the operator* (4.1) *be absolutely continuous. Then* $(D(\lambda))^{-1} \in H_\pm^{2,\mathrm{loc}}$, $(D(\lambda))^{-1}(\psi(t)(t-\lambda)^{-1}, \psi(t)) \in H_\pm^2$.

(ii) *Provided, that*

(a) $(D(\lambda))^{-1} \in H_\pm^{2+\delta,\mathrm{loc}}$, $\delta > 0$,

(b) $\psi(t) \in L_\infty(\mathbb{R})$,

the spectrum of the operator (4.1) *is absolutely continuous.*

The condition (3.9) for the operator (4.1) can now be reduced to the test of boundedness of certain singular integral operators acting in $L_2(\mathbb{R})$. Denote

$$\gamma(k) := \frac{1 - i\,\|\varphi\|\|\psi\|^{-1}(\psi(x)(x-k-i0)^{-1}, \psi(x))}{D(k+i0)}$$

$$\gamma_*(k) := \frac{1 - i\,\|\psi\|\|\varphi\|^{-1}(\varphi(x)(x-k-i0)^{-1}, \varphi(x))}{D_*(k+i0)},$$

where $D_*(\lambda) \equiv \overline{D(\overline{\lambda})}$.

The following theorem follows immediately:

Theorem 4.2. *Let the spectrum of the one-dimensional perturbation of the multiplication operator* (4.1) *be absolutely continuous. Assume that* $(\varphi, \psi) = 0$. *Assume that* $\gamma\varphi, \gamma_*\psi \in L_2(\delta)$ *for a given Borel set* $\delta \subset \mathbb{R}$. *Let singular integral operators defined by their respective kernels*

$$\begin{aligned} T_1(k,t) &= \frac{i\,\overline{\psi(t)}}{t-k} + \gamma(k)\frac{\overline{\varphi(t)}}{t-k}, \\ T_2(k,t) &= \frac{i\,\overline{\varphi(t)}}{t-k} + \gamma_*(k)\frac{\overline{\psi(t)}}{t-k}, \end{aligned} \tag{4.2}$$

i.e., the linear operators

$$T_1 : u \mapsto v.p. \int T_1(k,t)u(t)dt \text{ and } T_2 : u \mapsto v.p. \int T_2(k,t)u(t)dt$$

be bounded as operators acting from $L_2(\mathbb{R})$ *to* $L_2(\delta)$. *Then the restriction of the operator* L *to its spectral invariant subspace, corresponding to the set* $\delta \subset \mathbb{R}$, *is similar to a self-adjoint operator acting in* clos $\mathcal{P}_\delta N_e$.

The proof of this theorem is a straightforward application of the Corollary 3.7 to the operator under investigation. If one assumes some additional smoothness of the functions φ and ψ, the latter result can be formulated in a more concise form.

Corollary 4.3. *Let the spectrum of the one-dimensional perturbation of the multiplication operator* (4.1) *be absolutely continuous. Assume that* $(\varphi, \psi) = 0$ *and let further* $\varphi, \psi \in C^\beta(\mathbb{R})$ *for some* $\beta > 0$. *If singular integral operators* T_1' *and* T_2' *defined as above by their respective kernels*

$$\begin{aligned} T_1'(k,t) &= \frac{1}{D(k-i0)}\frac{\overline{\varphi(t)}}{t-k}, \\ T_2'(k,t) &= \frac{1}{D_*(k-i0)}\frac{\overline{\psi(t)}}{t-k} \end{aligned} \tag{4.3}$$

are bounded from $L_2(\mathbb{R})$ *to* $L_2(\delta)$ *for a given Borel set* $\delta \subset \mathbb{R}$, *then the restriction of the operator* L *to its spectral invariant subspace, corresponding to the set* $\delta \subset \mathbb{R}$, *is similar to a self-adjoint operator acting in* clos $\mathcal{P}_\delta N_e$.

Remark 4.4. The last result shows that under the assumption of some minimal additional smoothness the similarity problem reduces to the problem of (local) boundedness of a Hilbert-like transform in a weighted L_2, where the weight is determined by the determinant of perturbation D.

Acknowledgement

The author is grateful to Prof. S.N. Naboko for his constant attention to his work and to Prof. A. Sobolev for fruitful discussions during the author's stay in UCL.

The author is grateful to the Dept. of Mathematics, University College London for hospitality.

The author is also grateful to the referee for a number of extremely helpful remarks that hopefully have helped to improve the quality and overall readability of the paper.

References

[1] Arov D.Z., *Scattering theory with dissipation of energy*, Dokl. Akad. Nauk SSSR 216 (1974), 713–716 (Russian). English transl.: Sov. Math., Dokl. 15 (1974), 848–854.

[2] Azizov T., Iokhvidov I., *Linear operators in spaces with an indefinite metric*, Wiley-Interscience Publ., John Wiley & Sons Ltd., Chichester, 1989.

[3] Brodskij M.S., *Triangular and Jordan representations of linear operators*, English transl. in Amer. Math. Soc., Providence, R.I., 1971.

[4] Van Casteren J., *Boundedness properties of resolvents and semigroups of operators*, Acta Sci. Math. Szeged., vol. 48 (1980), N 1–2.

[5] K. Hoffman, *Banach spaces of analytic functions*, Prentice-Hall, Englewood Cliffs, N.J., 1962.

[6] Tosio Kato, *Perturbation theory for linear operators*, Springer-Verlag, 1966.

[7] Kiselev A.V., Faddeev M.M., *The similarity problem for non-self-adjoint operators with absolutely continuous spectrum*, Func. Anal. App. vol. 34 (2000), no. 2, pp. 140–142.

[8] Kiselev A.V., *Some Spectral Properties of the Nonself-Adjoint Friedrichs Model Operator*, Math. Proc. of R. Irish Acad., 105A (2005), no. 2, 25–46.

[9] Kiselev A.V., *On the similarity problem for the nonself-adjoint extensions of symmetric operators*, Operator Theory: Adv. Appl. Vol. 186 (2008), pp. 267–283.

[10] Lax, P.D., Phillips, R.S., *Scattering theory*, Pure and Applied Mathematics, 26. Academic Press, Inc., Boston, MA, 1989.

[11] Malamud M.M., *On the similarity of a triangular operator to a diagonal operator*, J. Math. Sci. (N.Y.) Vol. 115 (2003), no. 2, 2199–2222.

[12] Naboko S.N., *Absolutely continuous spectrum of a nondissipative operator and functional model. I.*, Zap. Nauchn. Semin. Leningr. Otd. Mat. Inst. Steklova, 65 (1976), 90–102 (Russian). English transl.: J. Sov. Math. 16 (1981), 1109–1117.

[13] Naboko S.N., *Absolutely continuous spectrum of a nondissipative operator and functional model. II.*, Zap. Nauchn. Semin. Leningr. Otd. Mat. Inst. Steklova, 73 (1977), 118–135 (Russian). English transl.: J. Sov. Math. 34 (1986), 2090–2101.

[14] Naboko S.N., *A functional model of perturbation theory and its applications to the scattering theory*, Tr. Mat. Inst. Steklova 147 (1980), 86–114 (Russian). English transl.: Proc. Steklov Inst. Math. 147 (1981), 85–116.

[15] Naboko S.N., *The conditions for similarity to unitary and selfadjoint operators*, Functional Analysis and its applications, vol. 18 (1984), pp. 16–27.

[16] Béla Sz.-Nagy and Ciprian Foiaş, *Analyse harmonique des opérateurs de l'espace de Hilbert*, Masson, Paris and Akad. Kiadó, Budapest, 1967.

[17] Nikolski N.K., *Operators, Functions and Systems: An Easy Reading*, vol. I, II, AMS, 2002.

[18] Nikolski N., Treil S., *Linear resolvent growth of rank one perturbation of a unitary operator does not imply its similarity to a normal operator,* Journal d'Analyse Mathématique 87 (2002).

[19] Pavlov B.S., *Conditions for separation of the spectral components of a dissipative operator*, Math. USSR-Izv., vol. 39 (1975), pp. 123–148.

[20] Pavlov B.S., *Dilation theory and spectral analysis of nonselfadjoint differential operators*, AMS Transl., vol. 115 (1980), no. 2.

[21] Romanov R.V., *A remark on equivalence of weak and strong definitions of absolutely continuous subspace for nonself-adjoint operators*, in Operator Theory: Advances and Applications (Proceedings of the OTAMP'02 Conference), 154 (2004), 179–184.

[22] Ryzhov V.A., *Absolutely continuous subspace of a nonself-adjoint operator and the scattering theory*, PhD Thesis, St. Petersburg, 1994 (Russian).

[23] Sahnovič L.A., *Nonunitary operators with absolutely continuous spectrum*, Izv. Akad. Nauk SSSR Ser. Mat., 33 (1969), 52–64 (Russian). English transl.: Math. USSR, Izv. 3 (1969), 51–63.

[24] Tikhonov A.S., *An absolutely continuous spectrum and a scattering theory for operators with spectrum on a curve* Algebra i Analiz 7 (1995), no. 1, 200–220. English transl.: St. Petersburg Math. J., 7 (1996), no. 1, 169–184.

[25] Veselov V.F., *Spectral decompositions of nonself-adjoint operators with singular spectrum*, PhD Thesis, Leningrad, 1986 (Russian).

[26] Veselov V.F., Naboko S.N., *The determinant of the characteristic function and the singular spectrum of a nonself-adjoint operator*, Mat. Sb., N. Ser. 129(171) (1986), No.1, 20–39 (Russian). English transl.: Math. USSR, Sb. 57 (1987), 21–41.

Alexander V. Kiselev
Department of Mathematics
University College
London WC1E 6BT, UK
(on leave from:
Department of Higher Mathematics and Mathematical Physics
St. Petersburg State University
1 Ulianovskaya Street, St. Petersburg
St. Peterhoff 198504 Russia)
e-mail: alexander.v.kiselev@gmail.com

Operator Theory:
Advances and Applications, Vol. 214, 83–106
© 2011 Springer Basel AG

Generalized Eigenfunctions and Spectral Theory for Strongly Local Dirichlet Forms

Daniel Lenz, Peter Stollmann and Ivan Veselić

Abstract. We present an introduction to the framework of strongly local Dirichlet forms and discuss connections between the existence of certain generalized eigenfunctions and spectral properties within this framework. The range of applications is illustrated by a list of examples.

Mathematics Subject Classification (2000). 35P05, 81Q10.

Keywords. Dirichlet forms, weak solutions, spectral properties.

Introduction

There is a long history to the study of connections between the spectrum of a selfadjoint differential operator and properties of generalized solutions to the associated eigenvalue equation. In this context, the following two "meta theorems" have attracted particular attention:

- Positive generalized eigenfunctions exist for energies below the spectrum and the spectrum begins at the energy, where positive generalized eigenfunctions cease to exist.
- The spectrum is given by those energies, for which a (suitably) bounded generalized solution exists.

The first statement is sometimes discussed under the name of "Allegretto Piepenbrink theorem". The second statement is discussed under the heading of "Shnol theorem". Precise versions (and proofs) of these statements have been given in various contexts. It turns out that the framework of (strongly local) Dirichlet forms allows one to give a unified and structurally rather simple discussion of these two results. This has recently be shown in [45] (for the Allegretto Piepenbrink theorem) and in [23] for the Shnol type result, see the results on expansion in eigenfunctions in [22] as well. The mentioned framework includes a variety of operators among them Schrödinger operators, (uniform) elliptic operators on manifolds and (suitable) quantum graphs. Accordingly, the mentioned results have a rather broad applicability.

Our aim here is to discuss this approach to basic spectral theory via Dirichlet forms in a way that is accessible to the non-specialist. In this way, we will not only feature the Shnol Theorem and the Allegretto Piepenbrink theorems of [23, 45] but also hope to advertise the use of Dirichlet forms in spectral problems. For this reason we also conclude the paper with a discussion of various applications. The results in this paper are concerned with strongly local Dirichlet forms. A study of similar results for non-local Dirichlet forms (e.g., graphs) can be found in [34].

The organisation of this paper is as follows. We give a introduction into Dirichlet forms in Sections 1, 2 and 3. This introduction is aimed at a non-specialist. We then discuss a version of Allegretto-Piepenbrink Theorem in Section 4 and results related to Shnol's Theorem in Section 5. These sections contain sketches of ideas and proofs. Finally, we discuss applications in Section 6.

1. Strongly local Dirichlet forms

In this section we describe the set-up used throughout the paper. We refer to [35] as to the classical standard reference as well as [21, 28, 36, 47] for literature on Dirichlet forms. We treat real and complex function spaces at the same time and write $\mathbb{K}$ to denote either $\mathbb{R}$ or $\mathbb{C}$.

Throughout we will work with a locally compact, separable metric space X endowed with a positive Radon measure m with $\operatorname{supp} m = X$.

Dirichlet forms

The central object of our studies is a regular Dirichlet form $\mathcal{E}$ with domain $\mathcal{D}$ in $L^2(X)$ and the selfadjoint operator H_0 associated with $\mathcal{E}$. In order to precisely define these notions we recall the basic terminology of Dirichlet forms: Consider a dense subspace $\mathcal{D} \subset L^2(X,m)$ and a sesquilinear and non-negative map $\mathcal{E}\colon \mathcal{D} \times \mathcal{D} \to \mathbb{K}$ such that $\mathcal{D}$ is closed with respect to the energy norm $\|\cdot\|_{\mathcal{E}}$, given by

$$\|u\|_{\mathcal{E}}^2 = \mathcal{E}[u,u] + \|u\|_{L^2(X,m)}^2,$$

in which case one speaks of a *closed form* in $L^2(X,m)$. In the sequel we will write

$$\mathcal{E}[u] := \mathcal{E}[u,u].$$

The selfadjoint operator H_0 associated with $\mathcal{E}$ is then characterized by

$$D(H_0) \subset \mathcal{D} \text{ and } \mathcal{E}[f,v] = (H_0 f \mid v) \quad (f \in D(H_0), v \in \mathcal{D}).$$

Such a closed form is said to be a *Dirichlet form* if $\mathcal{D}$ is stable under certain pointwise operations; more precisely, $T : \mathbb{K} \to \mathbb{K}$ is called a *normal contraction* if $T(0) = 0$ and $|T(\xi) - T(\zeta)| \le |\xi - \zeta|$ for any $\xi, \zeta \in \mathbb{K}$ and we require that for any $u \in \mathcal{D}$ also

$$T \circ u \in \mathcal{D} \quad \text{and} \quad \mathcal{E}[T \circ u] \le \mathcal{E}[u].$$

In the real case, this condition is often replaced by equivalent but formally weaker statement involving $u \vee 0$ and $u \wedge 1$, see [35], Thm. 1.4.1 and [47], Section I.4.

A Dirichlet form is called *regular* if $\mathcal{D} \cap C_c(X)$ is large enough so that it is dense both in $(\mathcal{D}, \| \cdot \|_{\mathcal{E}})$ and $(C_c(X), \| \cdot \|_\infty)$, where $C_c(X)$ denotes the space of continuous functions with compact support.

Examples

Here, we discuss some examples showing the wide range of applicability of Dirichlet forms.

The Laplacian on Euclidean space. The Laplacian in Euclidean space is the typical example to be kept in mind. It is given by

$$H_0 = -\Delta \text{ on } L^2(\Omega), \quad \Omega \subset \mathbb{R}^d \text{ open},$$

in which case

$$\mathcal{D} = W_0^{1,2}(\Omega) \text{ and } \mathcal{E}[u, v] = \int_\Omega (\nabla u | \nabla v) dx.$$

Note that for differentiable contractions $T : \mathbb{R} \longrightarrow \mathbb{R}$ the chain rule easily gives the crucial Dirichlet form property for real-valued functions u as

$$\mathcal{E}(Tu) = \int_\Omega (\nabla Tu, \nabla Tu) dx = \int_\Omega |T'(u(x))|^2 (\nabla u, \nabla u) dx \leq \mathcal{E}(u)$$

as $|T'(z)| \leq 1$ for all $z \in \mathbb{C}$.

Uniform elliptic operators in Euclidean space: In the previous example we can allow for quite irregular coefficients of the differential operator. More precisely, let $\Omega \subset \mathbb{R}^d$ open and let A be a measurable map from Ω into the symmetric $d \times d$ matrices. Assume that there exist $c, C >$ such that the eigenvalues of $A(x)$ lie in $[c, C]$ for all $x \in \Omega$. Then, the form $\mathcal{E}_A$ defined on $W_0^{1,2}(\Omega)$ by

$$\mathcal{E}_A[u, v] = \int_\Omega (A(x) \nabla u | \nabla v) dx$$

is a regular Dirichlet form.

Laplace Beltrami and unifom elliptic operators on manifolds: The previous example can easily be generalized to Laplace Beltrami operators on Riemannian manifolds: Let M be a Riemannian manifold with metric tensor g and exterior derivative d. Then, the form

$$\mathcal{E}_c(u, v) := \int_M (du, dv) dx$$

defined for $u, v \in C_c^\infty(M)$ is closable. The closure is a Dirichlet form and its domain of definition is given by $W_0^{1,2}(M)$. The generator is the Laplace Beltrami operator. Again, we can allow for a measurable map A from M into the symmetric linear maps on the corresponding cotangent spaces with eigenvalues lying in some interval $[c, C]$ for $c, C > 0$ and obtain the Dirichlet form

$$\mathcal{E}_A(u, v) := \int_M (A\, du, dv) dx$$

defined on $W_0^{1,2}(M)$. These examples can be further generalized to allow for some subriemannian manifolds. We will not give details here.

Quantum graphs with Kirchhoff boundary conditions: This example has received attention in recent times. We refrain from giving details here but refer to the last section of the paper.

Capacity

The *capacity* is a set function that allows one to measure the size of sets in a way that is adapted to the form $\mathcal{E}$.

For $U \subset X$, U open, we define

$$\operatorname{cap}(U) := \inf\{\|v\|_{\mathcal{E}}^2 \mid v \in \mathcal{D}, \chi_U \leq v\},$$

where we set $(\inf \emptyset = \infty)$. For arbitrary $A \subset X$, we then set

$$\operatorname{cap}(A) := \inf\{\operatorname{cap}(U) \mid A \subset U\}$$

(see [35], p. 61f.). We say that a property holds *quasi-everywhere*, short *q.e.*, if it holds outside a set of capacity 0. A function $f : X \to \mathbb{K}$ is said to be *quasi-continuous*, *q.c.* for short, if, for any $\varepsilon > 0$ there is an open set $U \subset X$ with $\operatorname{cap}(U) \leq \varepsilon$ so that the restriction of f to $X \setminus U$ is continuous.

A fundamental result in the theory of Dirichlet forms says that every $u \in \mathcal{D}$ admits a q.c. representative $\tilde{u} \in u$ (recall that $u \in L^2(X, m)$ is an equivalence class of functions) and that two such q.c. representatives agree q.e. Moreover, for every Cauchy sequence (u_n) in $(\mathcal{D}, \|\cdot\|_{\mathcal{E}})$ there is a subsequence (u_{n_k}) such that the $(\tilde{u}_{n_k})$ converge q.e. (see [35], p. 64f).

Whenever we will write expressions containing pointwise evaluations of functions u in the future, we will assume that a quasi continuous representative has been chosen.

Strong locality and the energy measure

$\mathcal{E}$ is called *strongly local* if

$$\mathcal{E}[u, v] = 0$$

whenever u is constant a.s. on the support of v.

Every strongly local, regular Dirichlet form $\mathcal{E}$ can be represented in the form

$$\mathcal{E}[u, v] = \int_X d\Gamma(u, v)$$

where Γ is a nonnegative sesquilinear mapping from $\mathcal{D}\times\mathcal{D}$ to the set of $\mathbb{K}$-valued Radon measures on X. It is determined by

$$\int_X \phi\, d\Gamma(u,u) = \mathcal{E}[u,\phi u] - \frac{1}{2}\mathcal{E}[u^2,\phi]$$

for real-valued $u\in\mathcal{D}$, $\phi\in\mathcal{D}\cap C_c(X)$ and called *energy measure*; see also [21].

Obviously, all examples discussed in the preceding subsection are strongly local. In the case of the Laplacian in Euclidean space, the measure Γ is given by $(\nabla u|\nabla v)dx$ appearing above.

We discuss properties of the energy measure next (see, e.g., [21, 35, 67]).

The energy measure inherits strong locality from $\mathcal{E}$ viz $\chi_U d\Gamma(\eta,u)=0$ holds for any open $U\Subset X$ and any $\eta,u\in\mathcal{D}$ with η constant on U. This directly allows one to extend Γ to $\mathcal{D}_{\mathrm{loc}}$ defined as

$$\big\{u\in L^2_{\mathrm{loc}}\mid \text{for all compact } K\subset X \text{ there is } \phi\in\mathcal{D} \text{ s.t. } \phi=u \text{ } m\text{-a.e. on } K\big\},$$

We will denote this extension by Γ again. This extension is **strongly local** again, i.e., satisfies

$$\chi_U d\Gamma(\eta,u)=0,$$

for any open $U\Subset X$ and any $\eta,u\in\mathcal{D}_{\mathrm{loc}}$ with η constant on U. The set $\mathcal{D}$ is then given as the set of all $u\in\mathcal{D}_{\mathrm{loc}}$ with $\int 1d\Gamma(u)<\infty$. The energy measure satisfies the **Leibniz rule**,

$$d\Gamma(u\cdot v,w)=ud\Gamma(v,w)+vd\Gamma(u,w),$$

for all $u,v\in\mathcal{D}_{\mathrm{loc}}\cap L^\infty_{\mathrm{loc}}(X)$. (In fact strong locality of $\mathcal{E}$ is equivalent to the validity of the Leibniz rule for functions in $\mathcal{D}\cap L^\infty_{\mathrm{loc}}$.) The energy measure also satisfies the chain rule

$$d\Gamma(\eta(u),w)=\eta'(u)d\Gamma(u,w)$$

whenever $u,w\in\mathcal{D}_{\mathrm{loc}}\cap L^\infty_{\mathrm{loc}}$ are real-valued and η is continuously differentiable.

We write $d\Gamma(u):=d\Gamma(u,u)$ and note that the energy measure satisfies the **Cauchy-Schwarz inequality**:

$$\begin{aligned}\int_X |fg|d|\Gamma(u,v)| &\le \left(\int_X |f|^2 d\Gamma(u)\right)^{\frac{1}{2}}\left(\int_X |g|^2 d\Gamma(v)\right)^{\frac{1}{2}}\\ &\le \frac{1}{2}\int_X |f|^2 d\Gamma(u)+\frac{1}{2}\int_X |g|^2 d\Gamma(v)\end{aligned}$$

for all $u,v\in\mathcal{D}_{\mathrm{loc}}$ and $f,g: X\longrightarrow\mathbb{C}$ measurable.

Due to Leibniz rule the sets $\mathcal{D}$ and $\mathcal{D}_{\mathrm{loc}}$ resp. have certain closedness properties under multiplication. This is an interesting feature and we discuss it next.

It is not hard to see that any function in $u\in\mathcal{D}_{\mathrm{loc}}$ with compact support belongs in fact to $\mathcal{D}$ (as $\int d\Gamma(u)=\int_{\operatorname{supp} u} d\Gamma(u)<\infty$). More generally, localized versions of functions from $\mathcal{D}_{\mathrm{loc}}$ belong to $\mathcal{D}$. More precisely, the following holds [45].

Lemma 1.1.

(a) *Let* $\Psi \in \mathcal{D}_{\text{loc}} \cap L^{\infty}_{\text{loc}}(X)$ *and* $\varphi \in \mathcal{D} \cap L^{\infty}_{c}(X)$ *be given. Then,* $\varphi\Psi$ *belongs to* $\mathcal{D}$.

(b) *Let* $\Psi \in \mathcal{D}_{\text{loc}}$ *and* $\varphi \in \mathcal{D} \cap L^{\infty}_{c}(X)$ *be such that* $d\Gamma(\varphi) \leq C \cdot dm$*. Then,* $\varphi\Psi$ *belongs to* $\mathcal{D}$.

Part (a) of the lemma gives in particular that $\mathcal{D} \cap C_c(X) = \mathcal{D}_{\text{loc}} \cap C_c(X)$ and $\mathcal{D} \cap L^{\infty}_{c}(X) = \mathcal{D}_{\text{loc}} \cap L^{\infty}_{c}(X)$ are closed under multiplication.

In order to introduce weak solutions on open subsets U of X, we extend $\mathcal{E}$ to $\mathcal{D}_{\text{loc}}(U) \times \mathcal{D}_c(U)$: where,

$$\mathcal{D}_{\text{loc}}(U) := \{u \in L^2_{\text{loc}}(U) \mid \forall \text{compact } K \subset U \exists\, \phi \in \mathcal{D} \text{ s.t. } \phi = u \ m\text{-a.e. on } K\}$$

$$\mathcal{D}_c(U) := \{\varphi \in \mathcal{D} | \operatorname{supp} \varphi \text{ compact in } U\}.$$

For $u \in \mathcal{D}_{\text{loc}}(U), \varphi \in \mathcal{D}_c(U)$ we define

$$\mathcal{E}[u, \varphi] := \mathcal{E}[\eta u, \varphi].$$

Here, $\eta \in \mathcal{D} \cap C_c(U)$ is arbitrary with constant value 1 on the support of φ. This makes sense as the RHS does not depend on the particular choice of η by strong locality.

Obviously, also Γ extends to a mapping $\Gamma : \mathcal{D}_{\text{loc}}(U) \times \mathcal{D}_{\text{loc}}(U) \to \mathcal{M}_R(U)$.

The intrinsic metric, strict locality and cut-off functions

Using the energy measure one can define the *intrinsic metric*

$$\rho : X \times X \longrightarrow [0, \infty]$$

by

$$\rho(x, y) = \sup\{|u(x) - u(y)| \ |u \in \mathcal{D}_{\text{loc}} \cap C(X) \text{ and } d\Gamma(u) \leq dm\}$$

where the latter condition signifies that $\Gamma(u)$ is absolutely continuous with respect to m and the Radon-Nikodym derivative is bounded by 1 on X. Despite its name, in general, ρ need not be a metric. However, it is a pseudo metric viz it is symmetric, satisfies $\rho(x, x) = 0$ for all $x \in X$ and satisfies the triangle inequality.

We say that $\mathcal{E}$ is *strictly local* if ρ is a metric that induces the original topology on X.

Note that strict locality implies that X is connected, since otherwise points x, y in different connected components would give $\rho(x, y) = \infty$, as characteristic functions of connected components are continuous and have vanishing energy measure.

We denote the intrinsic balls by

$$B(x, r) := \{y \in X | \rho(x, y) \leq r\}.$$

An important consequence of strict locality is that the distance function $\rho_x(\cdot) := \rho(x, \cdot)$ itself is a function in $\mathcal{D}_{\text{loc}}$ with $d\Gamma(\rho_x) \leq dm$, see [67]. This easily extends to the fact that for every closed $E \subset X$ the function $\rho_E(x) := \inf\{\rho(x, y) | y \in E\}$ enjoys the same properties (see the Appendix of [23]). This has a very important consequence. Whenever $\zeta : \mathbb{R} \longrightarrow \mathbb{R}$ is continuously differentiable, and $\eta := \zeta \circ \rho_E$, then η belongs to $\mathcal{D}_{\text{loc}}$ and satisfies

$$d\Gamma(\eta) = (\zeta' \circ \rho_E)^2 d\Gamma(\rho_E) \leq (\zeta' \circ \rho_E)^2 dm. \tag{1.1}$$

For this reason a lot of good cut-off functions are around in our context. More explicitly we note the following lemma (Lemma 1.3 in [45], see [23] as well).

Lemma 1.2. *For any compact K in X there exists a $\varphi \in C_c(X) \cap \mathcal{D}$ with $\varphi \equiv 1$ on K, $\varphi \geq 0$ and $d\Gamma(\varphi) \leq C\,dm$ for some $C > 0$. If L is another compact set containing K in its interior, then φ can be chosen to have support in L.*

Irreducibility

We will now discuss a notion that will be crucial in the proof of the existence of positive weak solutions below the spectrum. In what follows, $\mathfrak{h}$ will denote a densely defined, closed semibounded form in $L^2(X)$ with domain $D(\mathfrak{h})$ and positivity preserving semigroup $(T_t; t \geq 0)$. We denote by H the associated operator. Actually, the cases of interest in this paper are the situation that $\mathfrak{h} = \mathcal{E}$ is a Dirichlet form as discussed above, or a measure perturbation thereof $\mathfrak{h} = \mathcal{E} + \nu$. Here it is assumed that the positive and negative part of the measure ν obey $\nu_+ \in \mathcal{M}_{R,0}, \nu_- \in \mathcal{M}_{R,1}$, where the classes $\mathcal{M}_{R,0}, \mathcal{M}_{R,1}$ are discussed in the next section.

We say that $\mathfrak{h}$ is *reducible*, if there is a measurable set $M \subset X$ such that M and its complement M^c are nontrivial (have positive measure) and $L^2(M)$ is a reducing subspace for M, i.e., $\mathbb{1}_M D(\mathfrak{h}) \subset D(\mathfrak{h})$, $\mathfrak{h}$ restricted to $\mathbb{1}_M D(\mathfrak{h})$ is a closed form and $\mathcal{E}(u,v) = \mathcal{E}(u\mathbb{1}_M, v\mathbb{1}_M) + \mathcal{E}(u\mathbb{1}_{M^c}, v\mathbb{1}_{M^c})$ for all u, v. If there is no such decomposition of $\mathfrak{h}$, the latter form is called *irreducible*. Note that reducibility can be rephrased in terms of the semigroup and the resolvent:

Theorem 1.3. *Let $\mathfrak{h}$ be as above. Then the following conditions are equivalent:*

(i) *$\mathfrak{h}$ is irreducible.*

(ii) *T_t is positivity improving, for every $t > 0$, i.e., $f \geq 0$ and $f \neq 0$ implies that $T_t f > 0$ a.e.*

(iii) *$(H + E)^{-1}$ is positivity improving for every $E < \inf \sigma(H)$.*

We refer to [56], XIII.12 and a forthcoming paper [46] for details.

It is quite easy to see that a disconnected space easily leads to reducible forms. The converse is not exactly right, but there are recent results that go far in this direction and characterize irreducibility, cf. [31].

Assumptions

For the convenience of the reader we gather in this section assumptions and notation used in the sequel.

We will exclusively deal with regular, strongly local Dirichlet forms $\mathcal{E}$. The corresponding energy measure is denoted by Γ. The associated intrinsic metric is denoted by ρ.

We always choose quasi continuous representatives for elements of $\mathcal{D}_{\text{loc}}$.

The strongly local Dirichlet form $\mathcal{E}$ is *strictly local* if ρ is a metric that induces the original topology on X. This condition can be slightly weakened. It suffices to assume that the pseudometric ρ induces the original topology on X (as for a given $x \in X$ one can then always restrict attention to the set of y with $\rho(x,y) < \infty$).

Later we will also encounter a growth assumption on the intrinsic metric.

(G) All intrinsic balls have finite volume with subexponential growth:

$$e^{-\alpha \cdot R} m(B(x,R)) \to 0 \text{ as } R \to \infty \text{ for all } x \in X, \alpha > 0.$$

Finally, we note that $\mathcal{E}$ is called *ultracontractive* if for each $t > 0$ the semigroup e^{-tH_0} gives a map from $L^2(X)$ to $L^\infty(X)$.

2. Measure perturbations

We will be dealing with Schrödinger type operators, i.e., perturbations $H = H_0 + V$, where H_0 is associated to a strictly local Dirichlet form and the function V is a suitable potential. In fact, we can even include measures as potentials. Here, we follow the approach from [64, 65]. Measure perturbations have been regarded by a number of authors in different contexts, see, e.g., [11, 37, 67] and the references there.

We denote by $\mathcal{M}_R(U)$ the signed Radon measures on the open subset U of X and by $\mathcal{M}_{R,0}(U)$ the subset of measures ν that do not charge sets of capacity 0, i.e., those measures with $\nu(B) = 0$ for every Borel set B with $\mathrm{cap}(B) = 0$. In case that $\nu = \nu_+ - \nu_- \in \mathcal{M}_{R,0}(X)$ we can define

$$\nu[u,v] = \int_X \tilde{u}\overline{\tilde{v}} d\nu \text{ for } u, v \in \mathcal{D} \text{ with } \tilde{u}, \tilde{v} \in L^2(X, \nu_+ + \nu_-).$$

Of course, a special instance of such measures is given by $\nu = V dm$ whenever V belongs to $L^1_{\mathrm{loc}}(X)$.

We have to rely upon more restrictive assumptions concerning the negative part ν_- of our measure perturbation. We write $\mathcal{M}_{R,1}$ for those measures $\nu \in \mathcal{M}_R(X)$ that are $\mathcal{E}$-bounded with bound less than one; i.e., measures ν for which there is a $\kappa < 1$ and a $c_\kappa \geq 0$ such that

$$\nu[u,u] \leq \kappa \mathcal{E}[u] + c_\kappa \|u\|^2.$$

The set $\mathcal{M}_{R,1}$ can easily be seen to be a subset of $\mathcal{M}_{R,0}$.

By the KLMN theorem (see [55], p. 167), the sum $\mathcal{E} + \nu$ given by $D(\mathcal{E} + \nu) = \{u \in \mathcal{D} \mid \tilde{u} \in L^2(X, \nu_+)\}$ is closed and densely defined (in fact $\mathcal{D} \cap C_c(X) \subset D(\mathcal{E} + \nu)$) for ν with $\nu_+ \in \mathcal{M}_{R,0}, \nu_- \in \mathcal{M}_{R,1}$. We denote the associated selfadjoint operator by $H_0 + \nu$. Note that $\mathcal{D} \cap L^\infty_c(X) \subset D(\mathcal{E} + \nu)$.

An important subclass of $\mathcal{M}_{R,0}$ with very nice properties of the associated operators is the *Kato class* and the *extended Kato class*. In the present framework it can be defined in the following way: For $\mu \in \mathcal{M}_0$ and $\alpha > 0$ we set

$$\Phi(\mu,\alpha) : C_c(X)_+ \to [0,\infty],$$
$$\Phi(\mu,\alpha)\varphi := \int_X ((H_0+\alpha)^{-1}\varphi)^{\sim} \; d\mu.$$

The extended Kato class is defined as

$$\hat{\mathcal{S}}_K := \{\mu \in \mathcal{M}_0 | \exists \alpha > 0 : \Phi(\mu,\alpha) \in L^1(X,m)'\}$$

and, for $\mu \in \hat{\mathcal{S}}_K$ and $\alpha > 0$,

$$c_\alpha(\mu) := \|\Phi(\mu,\alpha)\|_{L^\infty(X,m)} (= \|\Phi(\mu,\alpha)\|_{L^1(X,m)'}), c_{\text{Kato}}(\mu) := \inf_{\alpha>0} c_\alpha(\mu).$$

The Kato class is originally defined via the fundamental solution of the Laplace equation in the classical case. In our setting it consists of those measures μ with $c_{\text{Kato}}(\mu) = 0$.

As done in various papers, one can even allow for more singular measures, a direction we are not going to explore here.

As already discussed our measure perturbations preserve closability of the form. They preserve further properties. In fact, regularity is preserved in our context as well.

Theorem 2.1 ([45]). *Let $(\mathcal{E},\mathcal{D})$ be a strongly local, regular Dirichlet form. Let ν with $\nu_+ \in \mathcal{M}_{R,0}, \nu_- \in \mathcal{M}_{R,1}$ be given. Then, the perturbed form $(\mathcal{E}+\nu, D(\mathcal{E}+\nu))$ is regular as well.*

Measure perturbations also preserve irreducibility, as can be seen from the following result.

Theorem 2.2 ([46]). *Let $(\mathcal{E},\mathcal{D})$ be a strictly local, regular, irreducible Dirichlet form. Let ν with $\nu_+ \in \mathcal{M}_{R,0}, \nu_- \in \mathcal{M}_{R,1}$ be given. Then, the perturbed form $(\mathcal{E}+\nu, D(\mathcal{E}+\nu))$ is irreducible as well.*

3. Weak solutions

Our main aim is to relate properties of weak solutions or generalized eigenfunctions to spectral properties of $H_0+\mu$. The necessary notation concerning weak solutions is introduced in this section. Throughout this section we consider a strongly local, regular Dirichlet form, $(\mathcal{E},\mathcal{D})$ on X and denote by $\Gamma : \mathcal{D}_{\text{loc}} \times \mathcal{D}_{\text{loc}} \to \mathcal{M}(X)$ the associated energy measure. We will be concerned with weak solutions Φ of the equation

$$(H_0+V)\Phi = \lambda \cdot \Phi, \tag{3.1}$$

where H_0 is the operator associated with $\mathcal{E}$ and V is a real-valued, locally integrable potential. In fact, we will consider a somewhat more general framework, allowing for measures instead of functions, as presented in the previous section. Moreover, we stress the fact that (3.1) is formal in the sense that Φ is not assumed to be in the operator domain of neither H_0 nor V. Here are the details.

Recall that we could extend Γ to a measure-valued function on U. In the same way, we can extend $\nu[\cdot,\cdot]$, using that every $u \in \mathcal{D}_{\mathrm{loc}}(U)$ admits a quasi continuous version $\tilde{u}$.

Definition 3.1. Let $U \subset X$ be open and $\nu \in \mathcal{M}_{R,0}(U)$ be a signed Radon measure on U that charges no set of capacity zero. Let $\lambda \in \mathbb{R}$ and $\Phi \in L^2_{\mathrm{loc}}(U)$. We say that Φ is a *weak solution* of $(H_0 + \nu)\Phi = \lambda \cdot \Phi$ in U if:

(i) $\Phi \in \mathcal{D}_{\mathrm{loc}}(U)$,

(ii) $\tilde{\Phi} d\nu \in \mathcal{M}_R(U)$,

(iii) $\forall \varphi \in \mathcal{D} \cap C_c(U), \quad \mathcal{E}[\Phi, \varphi] + \int_U \varphi \tilde{\Phi} d\nu = \lambda \cdot (\Phi|\varphi)$.

If $V \in L^1_{\mathrm{loc}}(U)$ we say that Φ is a *weak solution* of $(H_0 + V)\Phi = \lambda \cdot \Phi$ in U if it is a weak solution of $(H_0 + \nu)\Phi = \lambda \cdot \Phi$ for $\nu = V dm$.

Next, we briefly discuss these assumptions.

Remark 3.2.

(1) If $\nu = V dm$ and $V \in L^2_{\mathrm{loc}}(U)$, then property (ii) of the definition above is satisfied.

(2) If $\Phi \in L^\infty_{\mathrm{loc}}(U)$ and $\nu \in \mathcal{M}_R(U)$ then (ii) of the definition above is satisfied.

(3) If $\nu \in \mathcal{M}_R(U)$ satisfies (ii) above then $\nu - E dm \in \mathcal{M}_R(U)$ satisfies (ii) as well and any weak solution of $(H_0 + \nu)\Phi = \lambda \cdot \Phi$ in U is a weak solution of $(H_0 + \nu - E dm)\Phi = 0$ in U. Thus it suffices to consider the case $\lambda = 0$.

(4) By regularity we can replace (iii) by $\mathcal{E}[\Phi, \varphi] + \int_U \varphi \tilde{\Phi} d\nu = \lambda \cdot (\Phi|\varphi)$ for all $\varphi \in \mathcal{D}_{\mathrm{loc}} \cap L^\infty_c(U)$ (see [45] for details).

4. Positive weak solutions and the infimum of the spectrum

Throughout this section we consider a strongly local, regular Dirichlet form, $(\mathcal{E}, \mathcal{D})$ on X and denote by $\Gamma : \mathcal{D}_{\mathrm{loc}} \times \mathcal{D}_{\mathrm{loc}} \to \mathcal{M}(X)$ the associated energy measure. The results discussed in this section are taken from [45] to which we refer for further details and proofs.

Ground state transform and consequences

We start with a theorem giving the so-called ground state transform in our general setting.

Theorem 4.1. *Let $(\mathcal{E}, \mathcal{D})$ be a regular, strictly local Dirichlet form, H_0 be the associated operator and ν a measure with $\nu_+ \in \mathcal{M}_{R,0}, \nu_- \in \mathcal{M}_{R,1}$. Suppose that Φ is a weak solution of $(H_0 + \nu)\Phi = \lambda \cdot \Phi$ in X with $\Phi > 0$ m-a.e. and $\Phi, \Phi^{-1} \in L^\infty_{\mathrm{loc}}(X)$. Then, for all $\varphi, \psi \in D(\mathcal{E} + \nu)$, the products $\varphi\Phi^{-1}, \psi\Phi^{-1}$ belong to $\mathcal{D}_{\mathrm{loc}}$ and the formula*

$$\mathcal{E}[\varphi, \psi] + \nu[\varphi, \psi] = \int_X \Phi^2 d\Gamma(\varphi\Phi^{-1}, \psi\Phi^{-1}) + \lambda \cdot (\varphi|\psi) \tag{4.1}$$

holds.

The proof of the theorem proceeds essentially in two steps. In the first step a local version of the theorem is proven for "smooth" u, v. In the second step this local version is then extended to the whole space. Note also that the conditions on Φ in the theorem imply that Φ^{-1} is in $\mathcal{D}_{\mathrm{loc}}$. As the local version may be of independent interest and has a very simple proof we include statement and proof next.

Theorem 4.2. *Let $(\mathcal{E}, \mathcal{D})$ be a regular, strictly local Dirichlet form, H_0 be the associated operator and $\nu \in \mathcal{M}_{R,0}(U)$. Suppose that Φ is a weak solution of $(H_0 + \nu)\Phi = \lambda \cdot \Phi$ in U with $\Phi > 0$ m-a.e. and $\Phi, \Phi^{-1} \in L^\infty_{\mathrm{loc}}(U)$. Then, for all $\varphi, \psi \in \mathcal{D} \cap L^\infty_c(U)$:*

$$\mathcal{E}[\varphi, \psi] + \nu[\varphi, \psi] = \int_U \Phi^2 d\Gamma(\varphi\Phi^{-1}, \psi\Phi^{-1}) + \lambda \cdot (\varphi|\psi).$$

Proof. For the proof we may assume $\lambda = 0$ without restriction. Without loss of generality we may also assume that φ and ψ are real-valued functions. We now evaluate the RHS of the above equation, using the following identity. The Leibniz rule implies that for arbitrary $w \in \mathcal{D}_{\mathrm{loc}}(U)$:

$$0 = d\Gamma(w, 1) = d\Gamma(w, \Phi\Phi^{-1}) = \Phi^{-1} d\Gamma(w, \Phi) + \Phi d\Gamma(w, \Phi^{-1}) \qquad (\bigstar)$$

Therefore, for $\varphi, \psi \in \mathcal{D} \cap C_c(X)$:

$$\begin{aligned}
\int_X \Phi^2 d\Gamma(\varphi\Phi^{-1}, \psi\Phi^{-1}) &= \int_X \Phi d\Gamma(\varphi, \psi\Phi^{-1}) + \int_X \Phi^2 \varphi d\Gamma(\Phi^{-1}, \psi\Phi^{-1}) \\
\text{(by symmetry)} \quad &= \int_X d\Gamma(\varphi, \psi) + \int_X \Phi\psi d\Gamma(\varphi, \Phi^{-1}) \\
&\quad + \int_X \Phi^2 \varphi d\Gamma(\psi\Phi^{-1}, \Phi^{-1}) \\
&= \mathcal{E}[\varphi, \psi] + \int_X \Phi^2 d\Gamma(\varphi\psi\Phi^{-1}, \Phi^{-1}) \\
\text{(by } (\bigstar)) \quad &= \mathcal{E}[\varphi, \psi] - \int_X d\Gamma(\varphi\psi\Phi^{-1}, \Phi) \\
&= \mathcal{E}[\varphi, \psi] - \mathcal{E}[\varphi\psi\Phi^{-1}, \Phi].
\end{aligned}$$

As Φ is a weak solution we can now use part (4) of the previous remark to continue the computation by

$$\begin{aligned}
\cdots \quad &= \mathcal{E}[\varphi, \psi] - \big(-\nu[\varphi\psi\Phi^{-1}, \Phi]\big) \\
&= \mathcal{E}[\varphi, \psi] + \nu[\varphi, \psi].
\end{aligned}$$

This finishes the proof. □

The ideas behind our proof allow for some further generalizations. This is shortly indicated in the following remark.

Remark.

(1) This proof actually shows the following statement: Assume that there is a weak supersolution Φ of $(H_0 + \nu)\Phi = \lambda \cdot \Phi$ on X with $\Phi > 0$ m-a.e. and $\Phi, \Phi^{-1} \in L^\infty_{\text{loc}}(X)$. Then $\mathcal{E} + \nu \geq \lambda$.
(2) We can allow for complex measures ν without problems. In the context of PT-symmetric operators there is recent interest in this type of Schrödinger operators, see [15]
(3) Instead of measures also certain distributions could be included. Cf. [38] for such singular perturbations.

We explicitly note the following immediate consequence of (both) of the theorems of this section.

Corollary 4.3. *Let $(\mathcal{E}, \mathcal{D})$ be a regular, strictly local Dirichlet form, H_0 be the associated operator and ν a measure on X with $\nu_+ \in \mathcal{M}_{R,0}, \nu_- \in \mathcal{M}_{R,1}$. Suppose that Φ is a weak solution of $(H_0 + \nu)\Phi = \lambda \cdot \Phi$ in X with $\Phi > 0$ m-a.e. and $\Phi, \Phi^{-1} \in L^\infty_{\text{loc}}(X)$. Then, $H_0 + \nu \geq \lambda$.*

Harnack principles and existence of positive solutions below the spectrum

The previous subsection shows that $H_0 + \nu \geq \lambda$ whenever $\mathcal{E} + \nu$ is closable and admits a positive weak solution of $(H_0 + \nu)\Phi = \lambda\Phi$. In this subsection we discuss the converse under suitable conditions. A key property is related to the celebrated *Harnack inequality*.

Definition 4.4.

(1) We say that $H_0 + \nu$ *satisfies a Harnack inequality* for $\lambda \in \mathbb{R}$ if, for every relatively compact, connected open $X_0 \subset X$ there is a constant C such that all positive weak solutions Φ of $(H_0 + \nu)\Phi = \lambda\Phi$ on X_0 are locally bounded and satisfy

$$\text{esssup}_{B(x,r)} u \leq C \text{essinf}_{B(x,r)} u,$$

for every $B(x, r) \subset X_0$ where esssup and essinf denote the essential supremum and infimum.
(2) We say that $H_0 + \nu$ satisfies the *Harnack principle* for $\lambda \in \mathbb{R}$ if for every relatively compact, connected open subset U of X and every sequence $(\Phi_n)_{n\in\mathbb{N}}$ of nonnegative solutions of $(H_0 + \nu)\Phi = \lambda \cdot \Phi$ in U the following implication holds: If, for some measurable subset $A \subset U$ of positive measure

$$\sup_{n\in\mathbb{N}} \|\Phi_n 1\!\!1_A\|_2 < \infty$$

then, for all compact $K \subset U$ also

$$\sup_{n\in\mathbb{N}} \|\Phi_n 1\!\!1_K\|_2 < \infty.$$

(3) We say that $H_0 + \nu$ satisfies the *uniform Harnack principle* if for every bounded interval $I \subset \mathbb{R}$, every relatively compact, connected open subset U of X and every sequence $(\Phi_n)_{n\in\mathbb{N}}$ of nonnegative solutions of $(H_0+\nu)\Phi = \lambda_n \cdot \Phi$

in U with $\lambda_n \in I$ the following implication holds: If, for some measurable subset $A \subset U$ of positive measure

$$\sup_{n\in\mathbb{N}} \|\Phi_n 1\!\!1_A\|_2 < \infty$$

then, for all compact $K \subset U$ also

$$\sup_{n\in\mathbb{N}} \|\Phi_n 1\!\!1_K\|_2 < \infty.$$

Note that validity of a Harnack principle implies that a nonnegative weak solution Φ must vanish identically if it vanishes on a set of positive measure (as $\Phi_n := n\Phi$ has vanishing L^2 norm on the set of positive measure in question). Note also that validity of an Harnack inequality extends from balls to compact sets by a standard chain of balls argument. This easily shows that $H_0 + \nu$ satisfies the Harnack principle for $\lambda \in \mathbb{R}$ if it obeys a Harnack inequality for $\lambda \in \mathbb{R}$. Therefore, many situations are known in which the Harnack principle is satisfied:

For $\nu \equiv 0$ and $\lambda = 0$ a Harnack inequality holds, whenever $\mathcal{E}$ satisfies a Poincaré and a volume doubling property; cf. [20] and the discussion there. The most general results for $H_0 = -\Delta$ in terms of the measures ν that are allowed seem to be found in [37]. The uniformity of the estimates from [37] immediately gives that the uniform Harnack principle is satisfied for Kato class measures. Of the enormous list of papers on Harnack's inequality, let us also mention [8, 18, 19, 26, 37, 39, 41, 49, 57, 58, 69, 70]

Apart from the Harnack principle there is a second property that will be important in the proof of existence of positive general eigensolutions at energies below the spectrum.

Definition 4.5. The form $\mathcal{E}$ satisfies the *local compactness property* if $D_0(U) := \overline{D \cap C_c(U)}^{\|\cdot\|_{\mathcal{E}}}$ is compactly embedded in $L^2(X)$ for every relatively compact open $U \subset X$.

In case of the classical Dirichlet form the local compactness property follows from Rellich's Theorem on compactness of the embedding of Sobolev spaces in L^2.

It turns out that the situation is somewhat different depending on whether X is compact or not. In both cases we will need the assumption of irreducibility in order to obtain solutions which are positive almost everywhere. This is clear as in the reducible case a nontrivial solution could still vanish on some "components".

We first get the case of compact X out of our way.

Theorem 4.6. *Let $(\mathcal{E}, \mathcal{D})$ be a regular, strictly local, irreducible Dirichlet form, H_0 be the associated operator and ν a measure with $\nu_+ \in \mathcal{M}_{R,0}, \nu_- \in \mathcal{M}_{R,1}$. Suppose that X is compact and $\mathcal{E}$ satisfies the local compactness property. Then, $H_0+\nu$ has compact resolvent. In particular, there exists a positive weak solution to $(H_0+\nu)\Phi = \lambda_0\Phi$ for $\lambda_0 := \inf \sigma(H_0+\nu)$. This solution is unique (up to a factor) and belongs to $L^2(X)$. If $H_0+\nu$ satisfies a Harnack principle, then λ_0 is the only value in $\mathbb{R}$ allowing for a positive weak solution.*

We can now state our result in the case of non-compact X.

Theorem 4.7. *Let $(\mathcal{E}, \mathcal{D})$ be a regular, strictly local, irreducible Dirichlet form, H_0 be the associated operator and ν with $\nu_+ \in \mathcal{M}_{R,0}, \nu_- \in \mathcal{M}_{R,1}$. Suppose that $\mathcal{E}$ satisfies the local compactness property and X is noncompact. Then, if $\lambda < \inf \sigma(H_0 + \nu)$ and $H_0 + \nu$ satisfies the Harnack principle for λ, there is an a.e. positive solution of $(H_0 + \nu)\Phi = \lambda\Phi$.*

That we have to assume that X is noncompact can easily be seen by looking at the Laplacian on a compact manifold. In that situation any positive weak solution must in fact be in L^2 due to the Harnack principle. Thus the corresponding energy must lie in the spectrum (see Theorem 4.6).

Characterizing the infimum of the spectrum

The previous results do not yet settle the existence of a positive weak solution for the groundstate energy $\inf \sigma(H_0 + \nu)$ in the noncompact case. The uniform Harnack principle settles this question:

Theorem 4.8. *Let $(\mathcal{E}, \mathcal{D})$ be a regular, strictly local, irreducible Dirichlet form, H_0 be the associated operator, ν with $\nu_+ \in \mathcal{M}_{R,0}, \nu_- \in \mathcal{M}_{R,1}$. Suppose that $\mathcal{E}$ satisfies the local compactness property and $H_0 + \nu$ satisfies the uniform Harnack principle. Then there is an a.e. positive weak solution of $(H_0 + \nu)\Phi = \lambda\Phi$ for $\lambda = \inf \sigma(H_0 + \nu)$.*

5. Weak solutions and spectrum

In this section we relate energies in the spectrum to energies for which (suitably bounded) weak solutions exist. The results are taken from [23]. The final characterization relies on [22] as well.

A Weyl type criterion

We include the following criterion for completeness. It is taken from [66], Lemma 1.4.4, see also [30], Lemma 4.1 for the same result in a slightly different formulation.

Proposition 5.1. *Let h be a closed, semibounded form and H the associated self-adjoint operator. Then the following assertions are equivalent:*

(i) $\lambda \in \sigma(H)$.

(ii) *There exists a sequence (u_n) in $\mathcal{D}(h)$ with $\|u_n\| \to 1$ and*
$$\sup_{v \in \mathcal{D}(h), \|v\|_h \leq 1} |(h - \lambda)[u_n, v]| \to 0,$$
for $n \to \infty$.

A Caccioppoli type inequality

In this section we prove a bound on the energy measure of a generalized eigenfunction on a set in terms of bounds on the eigenfunction on certain neighborhood of the set.

We need the following notation: For $E \subset X$ and $b > 0$ we define the b-neighborhood of E as

$$B_b(E) := \{y \in X : \rho(y, E) \leq b\}.$$

Theorem 5.2. *Let $\mathcal{E}$ be a strictly local regular Dirichlet form. Let $\mu_+ \in \mathcal{M}_0$ and $\mu_- \in \mathcal{M}_1$ be given. Let $\lambda_0 \in \mathbb{R}$ and $b_0 > 0$ be given. Then, there exists a $C = C(b_0, \lambda_0, \mu_-)$ such that for any generalized eigenfunction u to an eigenvalue $\lambda \leq \lambda_0$ of $H_0 + \mu$ the inequality*

$$\int_E d\Gamma(u) \leq \frac{C}{b^2} \int_{B_b(E)} |u|^2 dm$$

holds for any closed $E \subset X$ and any $0 < b \leq b_0$.

Remark. For compact E both sides in the above inequality are finite, for E merely closed, one or both sides might be infinite. In any case, it suffices to prove the compact case since Γ is a Radon measure.

The Caccioppoli inequality replaces the familiar commutator estimates that are used for Schrödinger operators.

A 1/2 Shnol type result: How suitably bounded solutions force spectrum

In this section, we first present an abstract Shnol type result. Unfortunately, we have to start with a disclaimer. In [23] we messed up the reference to Shnol's original result (as do many other authors). In fact, [59] is the correct citation but there are two more papers with quite similar titles [60, 61] and [59] does not appear in MathSciNet.

The latter article deals with Schrödinger operators on the half-line and says that for spectrally almost every $\lambda \in \mathbb{R}$ the solution on the eigenvalue problem is bounded by $const\, x^{\frac{1}{2}+\varepsilon}$ as $x \longrightarrow \infty$ and vice versa. By "the solution" we mean a solution with the prescribed boundary condition at 0 and such a solution always exists since we are dealing with ODE. In this section we show $\frac{1}{2}$ Shnol, even a little stronger: if a weak solution with suitable exponential bounds exist for a given energy, that energy is in the spectrum.

We need the following notation. For $E \subset X$ and $b > 0$ we define the inner b-collar of E as

$$C_b(E) := \{y \in E : \rho(y, E^c) \leq b\}.$$

Theorem 5.3. *Let $\mathcal{E}$ be a strictly local regular Dirichlet form. Let $\mu_+ \in \mathcal{M}_0$ and $\mu_- \in \mathcal{M}_1$ be given. Let $\lambda \in \mathbb{R}$ with generalized eigenfunction u be given. If there exists $b > 0$ and a sequence (E_n) of compact subsets of X with*

$$\frac{\|u\chi_{C_b(E_n)}\|}{\|u\chi_{E_n}\|} \longrightarrow 0, n \longrightarrow 0,$$

then λ belongs to $\sigma(H)$.

We will now specialize our considerations to subexponentially bounded eigenfunctions.

A function $J : [0,\infty) \longrightarrow [0,\infty)$ is said to be subexponentially bounded if for any $\alpha > 0$ there exists a $C_\alpha \geq 0$ with $J(r) \leq C_\alpha \exp(\alpha r)$ for all $r > 0$. A $\mathbb{K}$-valued function f on a pseudo metric space (X, ρ) with measure m is said to be subexponentially bounded if for some $x_0 \in X$ and $\omega(x) = \rho(x_0, x)$ the function $e^{-\alpha\omega}u$ belongs to $L^2(X, m)$ for any $\alpha > 0$. Recall that a strictly local regular Dirichlet form $\mathcal{E}$ gives rise to an intrinsic pseudo metric ρ and an associated pseudo metric space (X, ρ).

Theorem 5.4. *Let $\mathcal{E}$ be a strictly local regular Dirichlet form, $x_0 \in X$ arbitrary and $\omega(x) = \rho(x_0, x)$. Let $\mu_+ \in \mathcal{M}_0$ and $\mu_- \in \mathcal{M}_1$ be given. Let $u \neq 0$ be a subexponentially bounded generalized eigenfunction. Then, λ belongs to $\sigma(H)$.*

A 1/2 Shnol type result: How spectrum forces suitably bounded generalized eigenfunctions.

In the last subsection we have discussed that existence of suitably bounded weak solutions implies that an energy belongs to the spectrum. In this section we discuss a converse given in [22] that was known before for ordinary Schrödinger operators; see the literature cited in the monograph [16].

Recall that $\mathcal{E}$ is called *ultracontractive* if for each $t > 0$ the semigroup e^{-tH_0} gives a map from $L^2(X)$ to $L^\infty(X)$.

Theorem 5.5. *Let $\mathcal{E}$ be a strictly local, regular, ultracontractive Dirichlet form satisfying condition* (G). *Let $\mu = \mu_+ - \mu_-$ with $\mu_+ \in \mathcal{M}_0$ and $\mu_- \in \hat{\mathcal{S}}_K$ with $c_{\text{Kato}}(\mu) < 1$. Define $H := H_0 + \mu$. Then for spectrally a.e. $\lambda \in \sigma(H)$ there is a subexponentially bounded generalized eigenfunction $u \neq 0$ with $Hu = \lambda u$.*

Actually, as remarked in [22], one does arrive at generalized eigenfunctions with polynomial bounds if one assumes that the volume of balls grows polynomially as well.

A Shnol type result: Characterizing the spectrum by subexponentially bounded solutions

We can now put together the results of the preceding subsections and obtain a characterization of the spectrum via subexponentially bounded solutions.

Corollary 5.6. *Let $\mathcal{E}$ be a strictly local, regular, ultracontractive Dirichlet form satisfying* (G). *Let $\mu = \mu_+ - \mu_-$ with $\mu_+ \in \mathcal{M}_0$ and $\mu_- \in \hat{\mathcal{S}}_K$ with $c_{\text{Kato}}(\mu) < 1$. Define $H := H_0 + \mu$. Then the spectral measures of H are supported on*

$$\{\lambda \in \mathbb{R} | \exists \text{ subexponentially bounded } u \neq 0 \text{ with } Hu = \lambda u\}.$$

6. Examples and applications

Several different types of operators to which our results can be applied have already been mentioned in Section 1. This includes classical examples like Schrödinger operators and symmetric elliptic second-order differential operators on unbounded domains in $\mathbb{R}^d$. More generally, Laplace-Beltrami operators and rather general elliptic second-order differential operators on Riemannian manifolds fall also within this class. In this section we will discuss in some more detail two types of examples which have attracted attention more recently, namely singular interaction Hamiltonians and quantum graphs. Moreover, we discuss here applications of the ground state transformation.

Hamiltonians with singular interactions

Hamiltonians with singular interactions arise when the Laplacian is perturbed by a perturbation which is localized on a set of Lebesgue measure zero. Here we consider more specifically operators with an interaction supported on an orientable, compact sub-manifold $M \subset \mathbb{R}^d$ of class C^2 and codimension one. The manifold M may or may not have a boundary. In the sequel we follow roughly the exposition in [43]. For more background see [24] or Appendix K of [10].

The simplest type of Hamiltonian with a potential perturbation supported on M is formally given by

$$(H_{\alpha\sigma_M})f(x) := \big(-\Delta - \alpha \cdot \delta(x - M)\big)f(x)\,, \tag{6.1}$$

where $\alpha > 0$ is a coupling constant. To show that the operator $H_{\alpha\sigma_M}$ can be given a rigorous meaning we establish next that it falls into the framework outlined in Section 2.

For this purpose denote by ν_M the Dirac measure in $\mathbb{R}^d$ with support on M. This means that for any Borel set $G \subset \mathbb{R}^d$ we have $\nu_M(G) = s_{d-1}(G \cap M)$. Here s_{d-1} is the $d-1$-dimensional surface measure on M. From Theorem 4.1 in [24] we infer that the measure ν_M belongs to the Kato class. In particular, for such a measure and an arbitrary $a > 0$ there exists $b_a < \infty$ such that

$$\int_{\mathbb{R}^d} |\psi(x)|^2 \nu_M(dx) \le a\|\nabla\psi\|^2 + b_a\|\psi\|^2\,.$$

As mentioned in Section 2 this implies that the form $\mathcal{E}_{\alpha\nu_M} := \mathcal{E} + \alpha\nu_M$ is closed on the domain $\mathcal{D}$ and densely defined. The unique selfadjoint operator associated to $\mathcal{E}_{\alpha\nu_M}$ acing on $L^2(\mathbb{R}^d)$ will be denoted by $H_{\alpha\nu_M}$.

It is possible to define the operator $H_{\alpha\nu_M}$ by appropriate selfadjoint boundary conditions on M, cf. [24, 43]. To explain this more precisely we need some notation. Denote by $\mathrm{n}\colon M \to \mathbb{S}^d$ a global unit normal vectorfield on M. Denote by $D(\tilde{H}_{\alpha\nu_M})$ the set of functions

$$\psi \in C(\mathbb{R}^d) \cap W^{1,2}(\mathbb{R}^d) \cap C^\infty(\mathbb{R}^d \setminus M) \cap W^{2,2}(\mathbb{R}^d \setminus M)$$

which satisfy for all $x \in M$

$$\lim_{\epsilon\searrow 0} \frac{\psi(x+\epsilon\mathrm{n}(x)) - \psi(x)}{\epsilon} + \lim_{\epsilon\searrow 0} \frac{\psi(x-\epsilon\mathrm{n}(x)) - \psi(x)}{\epsilon} = -\alpha\,\psi(x)$$

Using Green's formula one concludes as in Remark 4.1 of [24] that the closure of $-\Delta$ with domain $D(\tilde{H}_{\alpha\nu_M})$ is the selfadjoint operator $H_{\alpha\nu_M}$.

Since the measure ν_M belongs to the Kato class and is supported on a compact set, the essential spectrum of $H_{\alpha\nu_M}$ equals $[0,\infty)$, cf. Theorem 3.2 in [24]. In space dimension two $H_{\alpha\nu_M}$ has nonempty discrete spectrum for any positive value of the coupling constant α. This can be seen using the proof of Corollary 11 in [25]. For higher dimensions there is a critical value $\alpha_c > 0$ such that there exists a negative eigenvalue if and only if $\alpha \geq \alpha_c$, cf. the discussion on page 20 of [33].

Quantum graphs

Quantum graphs are given in terms of a metric graph Γ and a Laplace (or more generally) Schrödinger operator H defined on the edges of Γ together with a set of (generalised) boundary conditions at the vertices which make H selfadjoint. To make this more precise we define the geometric structure of metric graphs, as well as the operators acting on the associated L^2-Hilbert space.

We start with the definition of a metric graph which is appropriate for our purposes.

Definition 6.1. Let V and E be countable sets, l_- a positive real, and $\mathcal{G}$ a map

$$\mathcal{G}\colon E \to V \times V \times [l_-, \infty], \quad e \mapsto (\iota(e), \tau(e), l_e).$$

Here $[l_-, \infty]$ means $[l_-, \infty) \cup \{+\infty\}$. We call the triple $\Gamma = (V, E, \mathcal{G})$ a metric graph, elements of $V = V(\Gamma)$ vertices, elements of $E = E(\Gamma)$ edges, $\iota(e)$ the initial vertex of e, $\tau(e)$ the terminal vertex of e and l_e the length of e. Both $\iota(e)$ and $\tau(e)$ are called endvertices of e, or incident to e. The number of edges incident to the vertex v is called the *degree of* v. We assume that the degree is finite for all vertices.

Note that the two endvertices of an edge are allowed to coincide and there may be multiple edges connecting two vertices. We let $X_e := \{e\} \times (0, l_e)$, $X = X_\Gamma = V \cup \bigcup_{e \in E} X_e$ and $\overline{X_e} := \{e\} \times [0, l_e]$. On the set X it is possible to define in a natural way the length of paths and, using this notion, also a metric, cf. Section 1 in [44].

Now we introduce the relevant Hilbert spaces on which the Laplace, respectively, Schrödinger operators will act. For $k \in \{0, 1, 2\}$ we set

$$W^{k,2}(E) := \bigoplus_{e \in E} W^{k,2}(0, l_e)$$

and for $W^{0,2}(E)$ we use the usual notation $L^2(E)$. Given $k \in \{0, 1, 2\}$ and a function $u \in W^{1,k}(E)$ we denote by u_e the projection of u to the space $W^{k,2}(0, l_e)$. Thus we can identify each $u \in W^{1,k}(E)$ with a family $(u_e)_{e \in E}, u_e \in W^{0,2}(0, l_e)$.

Next we discuss pointwise properties of functions in $u \in W^{1,k}(0, l_e)$. Recall that for any $l > 0$ any element h of $W^{1,2}(0, l)$ has a continuous version; we will always pick this version and then the boundary value $h(0) := \lim_{x \to 0+} h(x)$ exists and satisfies

$$|h(0)|^2 \leq \frac{2}{l} \|h\|^2_{L^2(0,l)} + l \|h'\|^2_{L^2(0,l)} \tag{6.2}$$

by standard Sobolev type theorems. Consider now an edge e, the vertex $v = \iota(e) \in V$ and $u \in W^{1,2}(0, l_e)$. Then the limit $u(v) := \lim_{t \to 0} u(t)$ exists, as well as $u(w) := \lim_{t\to l_e} u(t)$ for $w = \tau(e)$ and (6.2) holds (with the obvious modifications). Similarly, for an edge e and the vertex $v = \iota(e)$ and the vertex $w = \tau(e)$ and $u \in W^{2,2}(0, l_e)$ the limits $u'(v) := \lim_{x\to v, x\in e} u'(x)$ and $u'(w) := -\lim_{x\to w, x\in e} u'(x)$ exist. Note that our sign convention is such that the definition of the derivative is canonical, i.e., independent of the choice of orientation of the edge. For $f \in W^{1,2}(E)$ and each vertex v we gather the boundary values of $f_e(v)$ over all edges e adjacent to v in a vector $f(v)$. More precisely, denote by $E_v := \{e \in E | v \in \{\iota(e), \tau(e)\}\}$ the set of edges incident to v and define $f(v) := (f_e(v))_{e\in E_v} \in \mathbb{C}^{E_v}$ and similarly, for $f \in W^{2,2}(E)$ we further collect the boundary values of $f'_e(v)$ over all edges e adjacent to v in a vector $f'(v) \in \mathbb{C}^{E_v}$. These boundary values of functions will be used to define the boundary conditions of the Laplacian, respectively the domains of definition of the forms we will be considering. Here we restrict ourselves to Kirchhoff boundary conditions and call a function $(u_e)_{e\in E} \in W^{1,2}(E)$ continuous, if, for any vertex v and all edges adjacent to it $u_e(v) = u_{e'}(v)$. Now set

$$D(s_0) := W^{1,2}(E) \cap C(X) \tag{6.3}$$

$$s_0(f, g) := \sum_{e\in E} \int_0^{l(e)} f'_e(t)\overline{g}'_e(t)dt \tag{6.4}$$

Obviously, the form s_0 is bounded below, closed a Dirichlet form and strongly local. Hence, there exists a unique associated self-adjoint operator which we denote by H_P. It can be explicitly characterized by

$$D(H_K) := \left\{ f \in W^{2,2}(E) \cap C(X) : \sum_{e\in E_v} f_e(v) = 0 \text{ for all } v \in V \right\}$$

$$(H_K f)_e : = -f''_e \text{ for all } e \in E.$$

It is possible to define quantum graphs with more general generalised boundary conditions at the vertices but not all reasonable choices will lead to Dirichlet forms; in [40] a characterization of those boundary conditions for which the form is a Dirichlet form is given. However the setup is somewhat different from ours.

Applications

The ground state transformation which featured in Theorem 4.1 can be used to obtain a formula for the lowest spectral gap. To be more precise let us assume that $\mathcal{E}$, ν and Φ satisfy the conditions of Theorem 4.1. Assume in addition that Φ is in $\mathcal{D}(\mathcal{E} + \nu)$. Then Φ is an eigenfunction of H corresponding to the eigenvalue $\lambda = \min \sigma(H)$. We denote by

$$\lambda' := \inf\{\mathcal{E}[u, u] + \nu[u, u] \mid u \in \mathcal{D}, \|u\| = 1, u \perp \Phi\}$$

the second lowest eigenvalue below the essential spectrum of H, or, if it does not exist, the bottom of $\sigma_{\text{ess}}(H)$. Then we obtain the following formula

$$\lambda' - \lambda = \inf_{\{u\in\mathcal{D}(\mathcal{E}+\nu), \|u\|=1, u\perp\Phi\}} \int_X \Phi^2 d\Gamma(u\Phi^{-1}, u\Phi^{-1}) \tag{6.5}$$

which determines the lowest spectral gap. It has been used in [42, 43, 72] to derive lower bounds on the distance between the two lowest eigenvalues of different classes of Schrödinger operators (see [63] for a related approach). In [42] bounded potentials are considered, in [43] singular interactions along curves in $\mathbb{R}^2$ are studied, and [72] generalizes these results using a unified approach based on Kato-class measures.

If for a subset $U \subset X$ of positive measure and a function $u \in \mathcal{D}$ with $\|u\| = 1$ and $u \perp \Phi$ the non-negative measure $\Gamma(u\Phi^{-1}, u\Phi^{-1})$ is absolutely continuous on U with respect to m, one can exploit formula (6.5) to derive the following estimate (cf. Section 3 in [72], and [42, 43] for similar bounds). Denote by $\gamma(u\Phi^{-1}) = \frac{d\Gamma(u\Phi^{-1}, u\Phi^{-1})}{dm}$ the Radon-Nykodim derivative. Then an application of the Cauchy-Schwarz inequality gives

$$\int_U \Phi^2 d\Gamma(u\Phi^{-1}, u\Phi^{-1}) \geq \frac{1}{m(U)} \inf_U \Phi^2 \left(\int_U \sqrt{\gamma(u\Phi^{-1})}dm \right)^2$$

Now we formulate more precisely the setting in which the above-mentioned results [42, 43, 72] apply. In fact, we choose here to formulate the main theorem of [72]. It applies to more general situations than [42] and [43] and is formulated in the language of Dirichlet forms. Consider the case where $X = \mathbb{R}^d$, $\mathcal{E}$ is equal to the classical Dirichlet form, ν is a non-negative, compactly supported measure satisfying for some $c_\nu \in (0, \infty), \alpha \in [0, 2)$ the bound $\nu(B(x, r)) \leq c_\nu r^{d-\alpha}$ for all $r > 0, x \in \mathbb{R}^d$, and D denotes the diameter of the support of ν. Let us assume that the bottom of the spectrum of $\mathcal{E} + \nu$ consists of two isolated eigenvalues, which will be denoted by $\lambda_0 < \lambda_1$. Under these assumptions there exist constants $C, C_0, p, q \in (0, \infty)$ such that

$$\lambda_1 - \lambda_0 \geq \frac{C}{(c_\nu + 1)^p (D + 1)^q} \cdot |\lambda_0| \cdot e^{-C_0 (D+1) \cdot \sqrt{|\lambda_0|}}$$

The ground state transformation plays an important role in other situations as well. It is for instance used in the study of L^p-L^q mapping properties of the semigroup associated to $\mathcal{E}$ [29]. In the theory of random Schrödinger operators it is used to remove a symmetry condition from the proof of of Lifschitz tails [48].

References

[1] S. Agmon. Lower bounds for solutions of Schrödinger equations. *J. Anal. Math.*, 23:1–25, 1970.

[2] S. Agmon. *On Positive Solutions of Elliptic Equations with Periodic Coefficients in R^N, Spectral Results and Extensions to Elliptic Operators on Riemannian Manifolds in Differential Equations* North Holland, Amsterdam, 1984.

[3] S. Agmon and L. Hörmander. Asymptotic properties of solutions of differential equations with simple characteristics. *J. d'Analyse Math.*, 30:1–38, 1976.

[4] S. Agmon. *Lectures on elliptic boundary value problems.* Van Nostrand Mathematical Studies, No. 2. Van Nostrand, Princeton, 1965.

[5] S. Agmon. *Lectures on exponential decay of solutions of second-order elliptic equations: bounds on eigenfunctions of N-body Schrödinger operators.* Princeton University Press, Princeton, N.J., 1982.

[6] S. Agmon. Bounds on exponential decay of eigenfunctions of Schrödinger operators. In *Schrödinger operators (Como,* 1984), volume 1159 of *Lecture Notes in Math.*, pages 1–38. Springer, Berlin, 1985.

[7] S. Agmon. On positivity and decay of solutions of second order elliptic equations on Riemannian manifolds. In *Methods of functional analysis and theory of elliptic equations (Naples,* 1982), pages 19–52. Liguori, Naples, 1983.

[8] M. Aizenman and B. Simon. Brownian motion and Harnack inequality for Schrödinger operators. *Commun. Pure Appl. Math.*, 35:209–273, 1982.

[9] M. Aizenman, R. Sims, and S. Warzel. Absolutely continuous spectra of quantum tree graphs with weak disorder. *Comm. Math. Phys.*, 264(2):371–389, 2006.

[10] S. Albeverio, F. Gesztesy, R. Høegh-Krohn, and H. Holden. *Solvable models in quantum mechanics.* AMS Chelsea Publishing, Providence, RI, second edition, 2005. With an appendix by P. Exner.

[11] S. Albeverio and Z.M. Ma. Perturbation of Dirichlet forms – lower semiboundedness, closability, and form cores. *J. Funct. Anal.*, 99(2):332–356, 1991.

[12] W. Allegretto. On the equivalence of two types of oscillation for elliptic operators. *Pac. J. Math.*, 55:319–328, 1974.

[13] W. Allegretto. Spectral estimates and oscillation of singular differential operators. *Proc. Am. Math. Soc.*, 73:51, 1979.

[14] W. Allegretto. Positive solutions and spectral properties of second order elliptic operators. *Pac. J. Math.*, 92:15–25, 1981.

[15] C.M. Bender, S. Boettcher and P. Meisinger. PT symmetric quantum mechanics. *J.Math.Phys.*, 40:2201–2229, 1999.

[16] Ju.M. Berezanskii. *Expansions in eigenfunctions of selfadjoint operators.* Translated from the Russian by R. Bolstein, J.M. Danskin, J. Rovnyak and L. Shulman. Translations of Mathematical Monographs, Vol. 17 American Mathematical Society, Providence, R.I. 1968

[17] A. Beurling and J. Deny. Dirichlet spaces. *Proc. Nat. Acad. Sci. U.S.A.*, 45:208–215, 1959.

[18] M. Biroli. Schrödinger type and relaxed Dirichlet problems for the subelliptic p-Laplacian. *Potential Analysis*, 15, 1–16, 2001.

[19] M. Biroli and S. Marchi. Harnack inequality for the Schrödinger problem relative to strongly local Riemannian p-homogeneous forms with a potential in the Kato class. *Bound. Value Probl.* 2007, Art. ID 24806, 19 pp.

[20] M. Biroli and U. Mosco. A Saint-Venant type principle for Dirichlet forms on discontinuous media. *Ann. Mat. Pura Appl., IV. Ser.*, 169:125–181, 1995.

[21] N. Bouleau and F. Hirsch. *Dirichlet forms and analysis on Wiener space*, volume 14 of *de Gruyter Studies in Mathematics.* Walter de Gruyter & Co., Berlin, 1991.

[22] A. Boutet de Monvel and P. Stollmann. Eigenfunction expansions for generators of Dirichlet forms. *J. Reine Angew. Math.*, 561:131–144, 2003.

[23] A. Boutet de Monvel, D. Lenz, and P. Stollmann. Schnol's theorem for strongly local forms. *Israel J. Math.*, to appear, 2008.

[24] J.F. Brasche, P. Exner, Y.A. Kuperin, and P. Šeba. Schrödinger operators with singular interactions. *J. Math. Anal. Appl.*, 184(1):112–139, 1994.

[25] J.F. Brasche. On eigenvalues and eigensolutions of the Schrödinger equation on the complement of a set with classical capacity zero *Methods Funct. Anal. Topology*, 9(3), 189–206, 2003. http://www.math.chalmers.se/ brasche/ett.pdf

[26] F. Chiarenza, E. Fabes, and N. Garofalo. Harnack's inequality for Schrödinger operators and the continuity of solutions. *Proc. Amer. Math. Soc.*, 98(3):415–425, 1986.

[27] H.L. Cycon, R.G. Froese, W. Kirsch, and B. Simon. *Schrödinger Operators with Application to Quantum Mechanics and Global Geometry.* Text and Monographs in Physics. Springer, Berlin, 1987.

[28] E.B. Davies. *Spectral theory and differential operators.* Cambridge University Press, Cambridge, 1995.

[29] E.B. Davies and B. Simon. Ultracontractivity and the heat kernel for Schrödinger operators and Dirichlet Laplacians. *J. Funct. Anal.*, 59(2):335–395, 1984.

[30] Y. Dermenjian, M. Durand and V. Iftimie: Spectral analysis of an acoustic multistratified perturbed cylinder. *Comm. Partial Differential Equations*, 23 (1-2): 141–169 (1998)

[31] A.F.M. ter Elst and D.W. Robinson: Invariant subspaces of submarkovian semigroups. *J. Evol. Equ.*, 8 (4): 661–671 (2008)

[32] P. Exner. An isoperimetric problem for leaky loops and related mean-chord inequalities. *J. Math. Phys.*, 46(6):062105, 2005. http://arxiv.org/abs/math-ph/0501066.

[33] P. Exner and K. Yoshitomi. Eigenvalue asymptotics for the Schrödinger operator with a δ-interaction on a punctured surface. *Lett. Math. Phys.*, 65(1): 19–26 (2003).

[34] R. Frank, D. Lenz, D. Wingert. Intrinsic metrics for non-local symmetric Dirichlet forms and applications. in preparation.

[35] M. Fukushima. *Dirichlet forms and Markov processes.* North-Holland Mathematical Library, Vol. 23. Amsterdam – Oxford – New York: North-Holland Publishing Company. Tokyo: Kodansha Ltd. X, 196 p., 1980.

[36] M. Fukushima, Y. Oshima, and M. Takeda. *Dirichlet forms and symmetric Markov processes.* de Gruyter Studies in Mathematics. 19. Berlin: Walter de Gruyter. viii, 392 p., 1994.

[37] W. Hansen. Harnack inequalities for Schrödinger operators. *Ann. Scuola Norm. Sup. Pisa Cl. Sci.* (4), 28(3):413–470, 1999.

[38] I.W. Herbst and A.D. Sloan. Perturbation of translation invariant positivity preserving semigroups on $L^2(\mathbf{R}^N)$. *Trans. Amer. Math. Soc.*, 236:325–360, 1978.

[39] M. Hoffmann-Ostenhof, T. Hoffmann-Ostenhof, and N. Nadirashvili. Interior Hölder estimates for solutions of Schrödinger equations and the regularity of nodal sets. *Commun. Part. Diff. Eqns*, 20:1241–1273, 1995.

[40] U. Kant, T. Klauß, J. Voigt and M. Weber. Dirichlet forms for singular one-dimensional operators and on graphs. *J. Evol. Equ.*, to appear.

[41] M. Kassmann. Harnack inequalities: An Introduction. *Bound. Value Probl.* 2007, Art. ID 81415, 21 pp. 35-01

[42] W. Kirsch and B. Simon. Comparison theorems for the gap of Schrödinger operators. *J. Funct. Anal.*, 75:396–410, 1987.

[43] S. Kondej and I. Veselić. Lower bounds on the lowest spectral gap of singular potential Hamiltonians. *Ann. Henri Poincaré*, 8(1):109–134, 2006.

[44] D. Lenz, C. Schubert, P. Stollmann. Eigenfunction expansion for Schrödinger operators on metric graphs. *Integral Equations and Operator Theory* 62 (2008), 541–553.

[45] D. Lenz, P. Stollmann and I. Veselić. The Allegretto-Piepenbrinck Theorem for strongly local Dirichlet forms. *Documenta Mathematica*, 14:167–190 (2009).

[46] D. Lenz, P. Stollmann and I. Veselić. Irreducibility and connectedness for Dirichlet forms. *In preparation*, 2009

[47] Z.-M. Ma and M. Röckner. *Introduction to the theory of (non-symmetric) Dirichlet forms.* Universitext. Berlin: Springer-Verlag, 1992.

[48] G.A. Mezincescu. Lifschitz singularities for periodic operators plus random potentials. *J. Statist. Phys.*, 49(5-6):1181–1190, 1987.

[49] J. Moser. On Harnacks theorem for elliptic differential equations. *Comm. Pure Appl. Math.*, 14:577–591, 1961.

[50] W.F. Moss and J. Piepenbrink. Positive solutions of elliptic equations. *Pacific J. Math.*, 75(1):219–226, 1978.

[51] J. Piepenbrink. Nonoscillatory elliptic equations. *J. Differential Equations*, 15:541–550, 1974.

[52] J. Piepenbrink. A conjecture of Glazman. *J. Differential Equations*, 24(2):173–177, 1977.

[53] Y. Pinchover. Topics in the theory of positive solutions of second-order elliptic and parabolic partial differential equations. In *Spectral theory and mathematical physics: a Festschrift in honor of Barry Simon's 60th birthday*, volume 76 of *Proc. Sympos. Pure Math.*, pages 329–355. Amer. Math. Soc., Providence, RI, 2007.

[54] R.G. Pinsky. *Positive harmonic functions and diffusion*, volume 45 of *Cambridge Studies in Advanced Mathematics.* Cambridge University Press, Cambridge, 1995.

[55] M. Reed and B. Simon. *Methods of Modern Mathematical Physics II, Fourier Analysis, Self-Adjointness.* Academic Press, San Diego, 1975.

[56] M. Reed and B. Simon. *Methods of Modern Mathematical Physics IV, Analysis of Operators.* Academic Press, San Diego, 1978.

[57] L. Saloff-Coste. Parabolic Harnack inequality for divergence-form second-order differential operators. *Potential Anal.*, 4(4):429–467, 1995. Potential theory and degenerate partial differential operators (Parma).

[58] J. Serrin. Local behavior of solutions of quasi-linear equations. *Acta Math.*, 111:247–302, 1964.

[59] È.È. Shnol. On the behaviour of eigenfunctions. (Russian) *Dokl. Akad. Nauk SSSR*, n. Ser. 94, 389–392 (1954).

[60] È.È. Shnol. The behaviour of eigenfunctions and the spectrum of Sturm–Liouville operators. (Russian) *Usp. Mat. Nauk* 9, No.4(62), 113–132 (1954)

[61] È.È. Shnol. On the behavior of the eigenfunctions of Schrödinger's equation. (Russian) *Mat. Sb. (N.S.)* 42 (84) (1957), 273–286; erratum 46 (88) (1958), 259.

[62] B. Simon. Schrödinger semigroups. *Bull. Amer. Math. Soc.* 7 (3): 447–526 (1982).

[63] I.M. Singer, B. Wong, S.-T. Yau, and S.S.-T. Yau. An estimate of the gap of the first two eigenvalues in the Schrödinger operator. *Ann. Scuola Norm. Sup. Pisa Cl. Sci.* (4), 12(2):319–333, 1985.

[64] P. Stollmann. Smooth perturbations of regular Dirichlet forms. *Proc. Am. Math. Soc.*, 116(3):747–752, 1992.

[65] P. Stollmann and J. Voigt. Perturbation of Dirichlet forms by measures. *Potential Anal.*, 5(2):109–138, 1996.

[66] P. Stollmann. *Caught by disorder: A Course on Bound States in Random Media*, volume 20 of *Progress in Mathematical Physics*. Birkhäuser, 2001.

[67] K.-T. Sturm. Analysis on local Dirichlet spaces. I: Recurrence, conservativeness and L^p-Liouville properties.

[68] K.-T. Sturm. Measures charging no polar sets and additive functionals of Brownian motion. *Forum Math.*, 4(3):257–297, 1992.

[69] K.-T. Sturm. Harnack's inequality for parabolic operators with singular low order terms. *Math. Z.*, 216(4):593–611, 1994.

[70] K.-T. Sturm. Analysis on local Dirichlet spaces. III: The parabolic Harnack inequality. *J. Math. Pures Appl.* (9) 75 (1996), 273–297.

[71] D. Sullivan. Related aspects of positivity in Riemannian geometry. *J. Differential Geom.*, 25(3):327–351, 1987.

[72] H. Vogt. A lower bound on the first spectral gap of Schrödinger operators with Kato class measures. *Ann. Henri Poincaré*, 10(2):395–414, (2009)

Daniel Lenz
Mathematisches Institut
Friedrich-Schiller Universität Jena
Ernst-Abbé Platz 2
D-07743 Jena, Germany
e-mail: `daniel.lenz@uni-jena.de`
URL: `http://www.analysis-lenz.uni-jena.de/`

Peter Stollmann
Fakultät für Mathematik
Technische Universität
D-09107 Chemnitz, Germany
e-mail: `peter.stollmann@mathematik.tu-chemnitz.de`

Ivan Veselić
Emmy-Noether-Project *Schrödingeroperatoren*
Fakultät für Mathematik
D-09107 TU Chemnitz, Germany
URL: `http://www.tu-chemnitz.de/mathematik/enp/`

Operator Theory:
Advances and Applications, Vol. 214, 107–124
© 2011 Springer Basel AG

Trace Formulas for Schrödinger Operators in Connection with Scattering Theory for Finite-gap Backgrounds

Alice Mikikits-Leitner and Gerald Teschl

Abstract. We investigate trace formulas for one-dimensional Schrödinger operators which are trace class perturbations of quasi-periodic finite-gap operators using Krein's spectral shift theory. In particular, we establish the conserved quantities for the solutions of the Korteweg–de Vries hierarchy in this class and relate them to the reflection coefficients via Abelian integrals on the underlying hyperelliptic Riemann surface.

Mathematics Subject Classification (2000). Primary 34L25, 35Q53; Secondary 81U40, 37K15.

Keywords. Scattering, KdV hierarchy, Trace formulas.

1. Introduction

Trace formulas for one-dimensional (discrete and continuous) Schrödinger operators have attracted an enormous amount of interest recently (see, e.g., [3], [18], [20], [27], [30], [34], [39]). However, most results are in connection with scattering theory for a constant background. On the other hand, scattering theory for one-dimensional Schrödinger operators with periodic background is a much older topic first investigated by Firsova in a series of papers [9]–[11]. Nevertheless, many questions which have long been answered in the constant background case are still open in this more general setting.

The aim of the present paper is to help filling some of these gaps. To this end, we want to find the analog of the classical trace formulas in scattering theory for the case of a quasi-periodic, finite-gap background. In the case of zero background it is well known that the transmission coefficient is the perturbation determinant in the sense of Krein [22] (see, e.g., [19], [32], [38] see also [15], [16] and the

Work supported by the Austrian Science Fund (FWF) under Grant No. Y330.

references therein for generalizations to non trace class situations) and our first aim is to establish this result for the case considered here; thereby establishing the connection with Krein's spectral shift theory. Our second aim is to find a representation of the transmission coefficient in terms of the scattering data – the analog of the classical Poisson–Jensen formula.

Moreover, scattering theory for one-dimensional Schrödinger operators is not only interesting in its own right, it also constitutes the main ingredient of the inverse scattering transform for the Korteweg–de Vries (KdV) hierarchy (see, e.g., [5], [28]). Again the case of decaying solutions is classical and trace formulas for this case were studied exhaustively in the past (cf. [14] and the references therein). Here we want to investigate the case of Schwartz type perturbations of a given finite-gap solution. The Cauchy problem for the KdV equation with initial conditions in this class was only solved recently by Egorova, Grunert, and Teschl [8] (see also [6], [7], [12]). Since the transmission coefficient is invariant when our Schrödinger operator evolves in time with respect to some equation of the KdV hierarchy, the corresponding trace formulas provide the conserved quantities for the KdV hierarchy in this setting.

Our work extends previous results for Jacobi operators by Michor and Teschl [29], [36]. For trace formulas in the pure finite-gap case see Gesztesy, Ratnaseelan, and Teschl [17] and Gesztesy and Holden [13].

2. Notation

We assume that the reader is familiar with quasi-periodic, finite-gap Schrödinger operators which arise naturally as the stationary solutions of the KdV hierarchy. Hence we only briefly recall some notation and refer to the monograph [13] (see also [28]) for further information.

Let

$$H_q = -\frac{d^2}{dx^2} + V_q(x) \tag{2.1}$$

be a finite-gap Schrödinger operator in $L^2(\mathbb{R})$ whose spectrum consists of $g+1$ bands:

$$\sigma(H_q) = \bigcup_{j=0}^{g-1} [E_{2j}, E_{2j+1}] \cup [E_{2g}, \infty). \tag{2.2}$$

It is well known that H_q is associated with the Riemann surface $\mathbb{M}$ of the function

$$R_{2g+1}^{1/2}(z), \qquad R_{2g+1}(z) = \prod_{j=0}^{2g} (z - E_j), \qquad E_0 < E_1 < \cdots < E_{2g}, \tag{2.3}$$

$g \in \mathbb{N}_0$. $\mathbb{M}$ is a compact, hyperelliptic Riemann surface of genus g. Here $R_{2g+1}^{1/2}(z)$ is chosen to have branch cuts along the spectrum with the sign fixed by the asymptotic behavior $R_{2g+1}^{1/2}(z) = \sqrt{z} z^g + \cdots$ as $z \to \infty$.

A point on $\mathbb{M}$ is denoted by $p = (z, \pm R_{2g+1}^{1/2}(z)) = (z, \pm)$, $z \in \mathbb{C}$. The point at infinity is denoted by $p_\infty = (\infty, \infty)$. We use $\pi(p) = z$ for the projection onto the extended complex plane $\mathbb{C} \cup \{\infty\}$. The points $\{(E_j, 0), 0 \le j \le 2g\} \cup \{p_\infty\} \subseteq \mathbb{M}$ are called branch points and the sets

$$\Pi_\pm = \{(z, \pm R_{2g+1}^{1/2}(z)) \mid z \in \mathbb{C}\backslash\Sigma\} \subset \mathbb{M}, \qquad \Sigma = \sigma(H_q), \tag{2.4}$$

are called upper and lower sheet, respectively. Note that the boundary of $\Pi_\pm$ consists of two copies of Σ corresponding to the two limits from the upper and lower half-plane.

We recall that upon fixing the spectrum $\sigma(H_q)$, the operator H_q is uniquely defined by choosing a Dirichlet divisor

$$\{(\mu_1, \sigma_1), \ldots, (\mu_g, \sigma_g)\}, \qquad \mu_j \in [E_{2j-1}, E_{2j}]. \tag{2.5}$$

For every $z \in \mathbb{C}$ the Baker–Akhiezer functions $\psi_{q,\pm}(z, x)$ are two (weak) solutions of $H_q \psi = z\psi$. They are the two branches of one function which is meromorphic on $\mathbb{M}\backslash\{p_\infty\}$ with simple poles at the Dirichlet divisor (2.5) and simple zeros at some other points

$$\{(\mu_1(x), \sigma_1(x)), \ldots, (\mu_g(x), \sigma_g(x))\}, \qquad \mu_j(x) \in [E_{2j-1}, E_{2j}], \tag{2.6}$$

which can be computed from the Dubrovin equations

$$\mu_j'(x) = \frac{-2\sigma_j(x) R_{2g+1}^{1/2}(\mu_j(x))}{\prod_{k \ne j} \mu_j(x) - \mu_k(x)} \tag{2.7}$$

using the initial conditions $\mu_j(0) = \mu_j$, $1 \le j \le g$. Moreover, $V_q(x)$ is explicitly given by the trace formula

$$V_q(x) = E_0 + \sum_{j=1}^{g} \big(E_{2j-1} + E_{2j} - 2\mu_j(x)\big). \tag{2.8}$$

The Baker–Akhiezer functions are linearly independent away from the band-edges $\{E_j\}_{j=0}^{2g}$ since their Wronskian is given by

$$W(\psi_{q,-}(z), \psi_{q,+}(z)) = \frac{2\mathrm{i} R_{2g+1}^{1/2}(z)}{\prod_{j=1}^{g}(z - \mu_j)}. \tag{2.9}$$

Here $W_x(f, g) = f(x)g'(x) - f'(x)g(x)$ denotes the usual Wronskian and μ_j are the Dirichlet eigenvalues at base point $x_0 = 0$. We recall that $\psi_{q,\pm}(z, x)$ have the form

$$\psi_{q,\pm}(z, x) = \theta_{q,\pm}(z, x) \exp(\pm \mathrm{i} x k(z)), \tag{2.10}$$

where $\theta_{q,\pm}(z, x)$ is quasi-periodic with respect to x and

$$k(z) = -\int_{E_0}^{p} \omega_{p_\infty,0}, \qquad p = (z, +), \tag{2.11}$$

denotes the quasimomentum map. Here $\omega_{p_\infty,k}$ is a normalized Abelian differential of the second kind with a single pole at $p_\infty = (\infty, \infty)$ and principal part $\zeta^{-k-2} d\zeta$

where $\zeta = z^{-1/2}$. It is explicitly given by

$$\omega_{p_\infty,0} = -\frac{\prod_{j=1}^{g}(\pi - \lambda_j)d\pi}{2R_{2g+1}^{1/2}}, \tag{2.12}$$

where $\lambda_j \in (E_{2j-1}, E_{2j})$, $1 \le j \le g$. In particular, $|\mathrm{e}^{\mathrm{i}k(z)}| < 1$ for $z \in \mathbb{C}\backslash\sigma(H_q)$ and $|\mathrm{e}^{\mathrm{i}k(z)}| = 1$ for $z \in \sigma(H_q)$.

3. Asymptotics of Jost solutions

After we have these preparations out of our way, we come to the study of short-range perturbations H of H_q associated with a potential V satisfying $V(x) \to V_q(x)$ as $|x| \to \infty$. More precisely, we will make the following assumption throughout this paper:

Let

$$H = -\frac{d^2}{dx^2} + V(x) \tag{3.1}$$

be a perturbation of H_q such that

$$\int_{-\infty}^{+\infty} |V(x) - V_q(x)| dx < \infty. \tag{3.2}$$

We first establish existence of Jost solutions, that is, solutions of the perturbed operator which asymptotically look like the Baker–Akhiezer solutions.

Theorem 3.1. *Assume* (3.2). *For every* $z \in \mathbb{C}\backslash\{E_j\}_{j=0}^{2g}$ *there exist* (*weak*) *solutions* $\psi_\pm(z,.)$ *of* $H\psi = z\psi$ *satisfying*

$$\lim_{x\to\pm\infty} \mathrm{e}^{\mp\mathrm{i}xk(z)}\big(\psi_\pm(z,x) - \psi_{q,\pm}(z,x)\big) = 0, \tag{3.3}$$

where $\psi_{q,\pm}(z,.)$ *are the Baker–Akhiezer functions. Moreover,* $\psi_\pm(z,.)$ *are continuous* (*resp. holomorphic*) *with respect to* z *whenever* $\psi_{q,\pm}(z,.)$ *are and*

$$\big|\mathrm{e}^{\mp\mathrm{i}xk(z)}\big(\psi_\pm(z,x) - \psi_{q,\pm}(z,x)\big)\big| \le C(z), \tag{3.4}$$

where $C(z)$ *denotes some constant depending only on* z.

Proof. Since $H\psi = z\psi$ is equivalent to $(H_q - z)\psi = -\widehat{V}\psi$, where $\widehat{V} = V - V_q$, we can use the variation of constants formula to obtain the usual Volterra integral equations for the Jost functions,

$$\begin{aligned}\psi_\pm(z,x) = \psi_{q,\pm}(z,x) - \frac{1}{W(\psi_{q,+},\psi_{q,-})}\int_x^{\pm\infty} &\big(\psi_{q,-}(z,x)\psi_{q,+}(z,y)\\ &- \psi_{q,-}(z,y)\psi_{q,+}(z,x)\big)\widehat{V}(y)\psi_\pm(z,y)dy.\end{aligned} \tag{3.5}$$

Moreover, introducing $\tilde{\psi}_\pm(z,x) = \mathrm{e}^{\mp\mathrm{i}xk(z)}\psi_\pm(z,x)$ the resulting integral equation can be solved using the method of successive iterations in the usual way. This proves the claims. □

Theorem 3.2. *Assume* (3.2). *The Jost functions have the following asymptotic behavior*

$$\psi_\pm(z,x) = \psi_{q,\pm}(z,x)\Big(1 \mp \frac{1}{2\mathrm{i}\sqrt{z}} \int_x^{\pm\infty} (V(y) - V_q(y))dy + o(z^{-1/2})\Big), \tag{3.6}$$

as $z \to \infty$, with the error being uniformly in x.

Proof. Invoking (3.5) we have

$$\begin{aligned}\frac{\psi_\pm(z,x)}{\psi_{q,\pm}(z,x)} &= 1 - \frac{1}{W(\psi_{q,+},\psi_{q,-})} \int_x^{\pm\infty} \Big(\psi_{q,-}(z,x)\psi_{q,+}(z,y)\frac{\psi_{q,\pm}(z,y)}{\psi_{q,\pm}(z,x)} \\ &\quad - \psi_{q,-}(z,y)\psi_{q,+}(z,x)\frac{\psi_{q,\pm}(z,y)}{\psi_{q,\pm}(z,x)}\Big)\widehat{V}(y)\frac{\psi_\pm(z,y)}{\psi_{q,\pm}(z,y)}dy \\ &= 1 \mp \int_x^{\pm\infty} \Big(G_q(z,x,x)\frac{\psi_{q,\pm}(z,y)^2}{\psi_{q,\pm}(z,x)^2} - G_q(z,y,y)\Big)\widehat{V}(y)\frac{\psi_\pm(z,y)}{\psi_{q,\pm}(z,y)}dy,\end{aligned} \tag{3.7}$$

where

$$G_q(z,x,y) = \frac{1}{W(\psi_{q,+},\psi_{q,-})}\begin{cases} \psi_{q,+}(z,x)\psi_{q,-}(z,y), & x \ge y, \\ \psi_{q,+}(z,y)\psi_{q,-}(z,x), & x \le y, \end{cases} \tag{3.8}$$

is the Green function of H_q. We have

$$G_q(z,x,x) = \frac{\psi_{q,+}(z,x)\psi_{q,-}(z,x)}{W(\psi_{q,+},\psi_{q,-})} = \frac{\mathrm{i}\prod_{j=1}^g (z - \mu_j(x))}{2R_{2g+1}^{1/2}(z)}. \tag{3.9}$$

Hence for z near ∞ one infers

$$G_q(z,x,x) = \frac{\mathrm{i}}{2\sqrt{z}}\Big(1 + \frac{1}{2}V_q(x)\frac{1}{z} + O\Big(\frac{1}{z^2}\Big)\Big), \tag{3.10}$$

where we made use of the trace formula (2.8). Next we insert (3.10) into (3.7) such that iteration implies

$$\frac{\psi_\pm(z,x)}{\psi_{q,\pm}(z,x)} = 1 \mp \frac{\mathrm{i}}{2\sqrt{z}}\Big(\int_x^{\pm\infty} \frac{\psi_{q,\pm}(z,y)^2}{\psi_{q,\pm}(z,x)^2}\widehat{V}(y)dy - \int_x^{\pm\infty} \widehat{V}(y)dy\Big) + O\left(\frac{1}{z}\right).$$

Next we will show that the first integral vanishes as $\sqrt{z} \to \infty$. We begin with the case $\mathrm{Im}(\sqrt{z}) \to \infty$. For that purpose note that

$$k(z) = \sqrt{z} + c + O(z^{-1/2}), \quad \text{as } z \to \infty,$$

for some constant $c \in \mathbb{C}$. Thus we compute

$$\begin{aligned}&\Big|\int_x^{\pm\infty} \frac{\psi_{q,\pm}(z,y)^2}{\psi_{q,\pm}(z,x)^2}\widehat{V}(y)dy\Big| \le C\int_x^{\pm\infty} \exp\big(\mp 2\mathrm{Im}(\sqrt{z})(y-x)\big)\big|\widehat{V}(y)\big|dy \\ &\le C\int_x^{x+\varepsilon} \big|\widehat{V}(y)\big|dy + C\cdot\exp\big(\mp 2\mathrm{Im}(\sqrt{z})\varepsilon\big)\int_{x+\varepsilon}^{\pm\infty} \big|\widehat{V}(y)\big|dy,\end{aligned}$$

such that the first integral can be made arbitrary small if $\varepsilon > 0$ is small and the second integral vanishes as $\mathrm{Im}(\sqrt{z}) \to \infty$.

Otherwise, if $\mathrm{Re}(\sqrt{z}) \to \infty$, we use (2.10) to rewrite the integral as

$$\int_x^{\pm\infty} \left(\frac{\theta_{q,\pm}(z,y)^2}{\theta_{q,\pm}(z,x)^2} \widehat{V}(y) \exp\big(\mp 2\mathrm{Im}(\sqrt{z})(y-x)\big) \right) \exp\big(\pm 2\mathrm{i}\mathrm{Re}(\sqrt{z})(y-x)\big) dy.$$

Since

$$\left| \frac{\theta_{q,\pm}(z,y)^2}{\theta_{q,\pm}(z,x)^2} \widehat{V}(y) \exp\big(\mp 2\mathrm{Im}(\sqrt{z})(y-x)\big) \right| \le C|\widehat{V}(y)|$$

the integral vanishes as $\mathrm{Re}(\sqrt{z}) \to \infty$ by a slight variation of the Riemann–Lebesgue lemma.

Hence we finally have

$$\frac{\psi_\pm(z,x)}{\psi_{q,\pm}(z,x)} = 1 \pm \frac{\mathrm{i}}{2\sqrt{z}} \int_x^{\pm\infty} \widehat{V}(y)dy + o\left(\frac{1}{\sqrt{z}}\right) \tag{3.11}$$

as $z \to \infty$. □

For later use we note the following immediate consequence

Corollary 3.3. *Under the assumptions of the previous theorem we have*

$$\lim_{x\to\pm\infty} \mathrm{e}^{\mp \mathrm{i}xk(z)} \Big(\dot{\psi}_\pm(z,x) \mp \mathrm{i}x\dot{k}(z)\psi_\pm(z,x) - \dot{\psi}_{q,\pm}(z,x) \pm \mathrm{i}x\dot{k}(z)\psi_{q,\pm}(z,x)\Big) = 0,$$

where the dot denotes differentiation with respect to z.

Proof. Just differentiate (3.3) with respect to z, which is permissible by uniform convergence on compact subsets of $\mathbb{C}\backslash\{E_j\}_{j=0}^{2g}$. □

We remark that if we require our perturbation to satisfy the usual short-range assumption as in [1], [9, 10, 11] (i.e., the first moment is integrable, see (5.1)), then we even have $\mathrm{e}^{\mp \mathrm{i}xk(z)}(\dot{\psi}_\pm(z,x) - \dot{\psi}_{q,\pm}(z,x)) \to 0$.

From Theorem 3.2 we obtain a complete characterization of the spectrum of H.

Theorem 3.4. *Assume* (3.2). *Then $(H-z)^{-1} - (H_q - z)^{-1}$ is trace class. In particular, we have $\sigma_{\mathrm{ess}}(H) = \sigma(H_q)$ and the point spectrum of H is confined to $\mathbb{R}\backslash\sigma(H_q)$. Furthermore, the essential spectrum of H is purely absolutely continuous except for possible eigenvalues at the band edges.*

Proof. That $(H-z)^{-1} - (H_q - z)^{-1}$ is trace class follows as in [37, Lem. 9.34] (cf. also [25, Sect. 4]). The fact that the essential spectrum is purely absolutely continuous follows from subordinacy theory ([37, Sect. 9.5]) since the asymptotics of the Jost solutions imply that no solution is subordinate inside the essential spectrum. □

Note that (3.2) does neither exclude eigenvalues at the boundary of the essential spectrum nor an infinite number of eigenvalues inside essential spectral gaps (see [31], [26] or [1] for conditions excluding these cases).

Our next result concerns the asymptotics of the Jost solutions at the *other side.*

Lemma 3.5. *Assume* (3.2). *Then the Jost solutions* $\psi_\pm(z,.)$, $z \in \mathbb{C}\backslash\sigma(H)$, *satisfy*

$$\lim_{x\to\mp\infty} \left|e^{\mp i x k(z)}\big(\psi_\pm(z,x) - \alpha(z)\psi_{q,\pm}(z,x)\big)\right| = 0, \tag{3.12}$$

where

$$\alpha(z) = \frac{W(\psi_-(z),\psi_+(z))}{W(\psi_{q,-}(z),\psi_{q,+}(z))}\frac{\prod_{j=1}^{g}(z-\mu_j)}{2iR_{2g+1}^{1/2}(z)}W(\psi_-(z),\psi_+(z)). \tag{3.13}$$

Proof. Since H and H_q have the same form domain, the second resolvent equation ([37, Lem. 6.30]) for form perturbations implies

$$G(z,x,x) - G_q(z,x,x) = \int_{-\infty}^{\infty} G(z,x,y)\widehat{V}(y)G_q(z,y,x)dy,$$

where $\widehat{V} = V - V_q$. By (3.8) and

$$G(z,x,y) = \frac{1}{W(\psi_+,\psi_-)}\begin{cases} \psi_+(z,x)\psi_-(z,y), & x \ge y,\\ \psi_+(z,y)\psi_-(z,x), & x \le y,\end{cases}$$

we obtain

$$\begin{aligned} G(z,x,x) - G_q(z,x,x) =& \frac{\psi_{q,+}(z,x)\psi_+(z,x)}{W(z)W_q(z)}\int_{-\infty}^{x}\widehat{V}(y)\psi_{q,-}(z,y)\psi_-(z,y)dy\\ &+ \frac{\psi_{q,-}(z,x)\psi_-(z,x)}{W(z)W_q(z)}\int_{x}^{\infty}\widehat{V}(y)\psi_{q,+}(z,y)\psi_+(z,y)dy, \end{aligned} \tag{3.14}$$

where $W(z) = W(\psi_+,\psi_-)$ and $W_q(z) = W(\psi_{q,+},\psi_{q,-})$. Next, by (3.4), note that

$$\left|\psi_{q,\pm}(z,x)\right| \le c_1 e^{\mp\varepsilon x}, \quad \left|\psi_\pm(z,x)\right| \le c_2 e^{\mp\varepsilon x}, \tag{3.15}$$

as $x \to +\infty$, where c_1, c_2 denote some constants and $\varepsilon > 0$ does only depend on z. Now one can show that the first term in (3.14) tends to 0 when $x \to +\infty$ using the same kind of argument as in the proof of Theorem 3.2. Similarly one then checks that the second term in (3.14) tends to 0 when $x \to -\infty$. Thus

$$\lim_{x\to\pm\infty} G(z,x,x) - G_q(z,x,x) = 0$$

and using

$$G_q(z,x,x) = \frac{\psi_{q,-}(z,x)\psi_{q,+}(z,x)}{W(\psi_{q,-}(z),\psi_{q,+}(z))}, \qquad G(z,x,x) = \frac{\psi_-(z,x)\psi_+(z,x)}{W(\psi_-(z),\psi_+(z))}$$

implies

$$\lim_{x\to\pm\infty}\big(\psi_-(z,x)\psi_+(z,x) - \alpha(z)\psi_{q,-}(z,x)\psi_{q,+}(z,x)\big) = 0,$$

respectively,

$$\lim_{x\to-\infty}\psi_{q,-}(z,x)\big(\psi_+(z,x) - \alpha(z)\psi_{q,+}(z,x)\big) = 0,$$

which is the claimed result. □

To see the connection with scattering theory (see, e.g., [1]), we introduce the scattering relations

$$T(\lambda)\psi_\pm(\lambda,x) = \overline{\psi_\mp(\lambda,x)} + R_\mp(\lambda)\psi_\mp(\lambda,x), \quad \lambda\in\sigma(H_q), \tag{3.16}$$

where the transmission and reflection coefficients are defined as usual,

$$T(\lambda) = \frac{W(\overline{\psi_\pm(\lambda)},\psi_\pm(\lambda))}{W(\psi_\mp(\lambda),\psi_\pm(\lambda))}, \qquad R_\pm(\lambda) := -\frac{W(\psi_\mp(\lambda),\overline{\psi_\pm(\lambda)})}{W(\psi_\mp(\lambda),\psi_\pm(\lambda))}, \quad \lambda\in\sigma(H_q). \tag{3.17}$$

In particular, $\alpha(z)$ is just the inverse of the transmission coefficient $T(z)$. It is holomorphic in $\mathbb{C}\backslash\sigma(H_q)$ with simple zeros at the discrete eigenvalues of H.

Corollary 3.6. *Assume* (3.2). *Then we have*

$$T(z) = \exp\Big(-\int_{-\infty}^{+\infty}(m_\pm(z,x) - m_{q,\pm}(z,x))dx\Big), \tag{3.18}$$

where

$$m_\pm(z,x) = \pm\frac{\psi_\pm'(z,x)}{\psi_\pm(z,x)}, \quad m_{q,\pm}(z,x) = \pm\frac{\psi_{q,\pm}'(z,x)}{\psi_{q,\pm}(z,x)} \tag{3.19}$$

are the Weyl–Titchmarsh functions. Here the prime denotes differentiation with respect to x.

Proof. From the definition (3.19) we get the following representations of the Jost and Baker–Akhiezer functions

$$\psi_\pm(z,x) = \psi_\pm(z,x_0)\exp\Big(\pm\int_{x_0}^x m_\pm(z,y)dy\Big),$$
$$\psi_{q,\pm}(z,x) = \psi_{q,\pm}(z,x_0)\exp\Big(\pm\int_{x_0}^x m_{q,\pm}(z,y)dy\Big),$$

and thus

$$\begin{aligned}\frac{\psi_\pm(z,x)}{\psi_{q,\pm}(z,x)} &= \frac{\psi_\pm(z,x_0)}{\psi_{q,\pm}(z,x_0)}\exp\Big(\pm\int_{x_0}^x(m_\pm(z,y)-m_{q,\pm}(z,y))dy\Big)\\ &= \exp\big(\pm\int_{\pm\infty}^x(m_\pm(z,y)-m_{q,\pm}(z,y))dy\big).\end{aligned}$$

Making use of that and (3.12) we get

$$\alpha(z) = \lim_{x\to\mp\infty}\frac{\psi_\pm(z,x)}{\psi_{q,\pm}(z,x)}\exp\Big(\pm\int_{\pm\infty}^{\mp\infty}(m_\pm(z,y)-m_{q,\pm}(z,y))dy\Big),$$

which finishes the proof. □

Corollary 3.7. *Assume* (3.2). *Then $T(z)$ has the following asymptotic behavior*

$$T(z) = 1 + \frac{1}{2\mathrm{i}\sqrt{z}}\int_{-\infty}^{\infty}(V(y)-V_q(y))dy + o(z^{-1/2}), \tag{3.20}$$

as $z\to\infty$.

Proof. Use (3.12) and (3.6). □

4. Connections with Krein's spectral shift theory and trace formulas

To establish the connection with Krein's spectral shift theory we next show:

Lemma 4.1. *We have*

$$\frac{d}{dz}\alpha(z) = -\alpha(z)\int_{-\infty}^{+\infty} (G(z,x,x) - G_q(z,x,x))dx, \qquad z \in \mathbb{C}\backslash\sigma(H), \tag{4.1}$$

where $G(z,x,y)$ *and* $G_q(z,x,y)$ *are the Green's functions of* H *and* H_q*, respectively.*

Proof. The Lagrange identity ([37], eq. (9.4)) implies

$$W_x(\psi_+(z),\dot{\psi}_-(z)) - W_y(\psi_+(z),\dot{\psi}_-(z)) = \int_y^x \psi_+(z,r)\psi_-(z,r)dr, \tag{4.2}$$

hence the derivative of the Wronskian can be written as

$$\begin{aligned}\frac{d}{dz}W(\psi_-(z),\psi_+(z)) &= W_x(\dot{\psi}_-(z),\psi_+(z)) + W_x(\psi_-(z),\dot{\psi}_+(z))\\ &= W_y(\dot{\psi}_-(z),\psi_+(z)) + W_x(\psi_-(z),\dot{\psi}_+(z)) - \int_y^x \psi_+(z,r)\psi_-(z,r)dr.\end{aligned}$$

Using Corollary 3.3 and Lemma 3.5 we have

$$\begin{aligned}W_y(\dot{\psi}_-(z),\psi_+(z)) &= W_y(\dot{\psi}_- + \mathrm{i}\dot{k}y\psi_-,\psi_+)\\ &\quad -\mathrm{i}\dot{k}\big(y\,W(\psi_-,\psi_+) - \psi_-(z,y)\psi_+(z,y)\big)\\ &\to \alpha W_y(\dot{\psi}_{q,-} + \mathrm{i}\dot{k}y\psi_{q,-},\psi_{q,+})\\ &\quad -\alpha\mathrm{i}\dot{k}\big(y\,W(\psi_{q,-},\psi_{q,+}) - \psi_{q,-}(z,y)\psi_{q,+}(z,y)\big)\\ &= \alpha(z)W_y(\dot{\psi}_{q,-}(z),\psi_{q,+}(z))\end{aligned}$$

as $y \to -\infty$. Similarly we obtain

$$W_x(\psi_-(z),\dot{\psi}_+(z)) \to \alpha(z)W_x(\psi_{q,-}(z),\dot{\psi}_{q,+}(z))$$

as $x \to +\infty$ and again using (4.2) we have

$$W_y(\dot{\psi}_{q,-}(z),\psi_{q,+}(z)) = W_x(\dot{\psi}_{q,-}(z),\psi_{q,+}(z)) + \int_y^x \psi_{q,+}(z,r)\psi_{q,-}(z,r)dr.$$

Collecting terms we arrive at

$$\begin{aligned}\dot{W}(\psi_-(z),\psi_+(z)) = &-\int_{-\infty}^{+\infty}\Big(\psi_+(z,r)\psi_-(z,r) - \alpha(z)\psi_{q,+}(z,r)\psi_{q,-}(z,r)\Big)dr\\ &+\alpha(z)\dot{W}(\psi_{q,-}(z)\psi_{q,+}(z)).\end{aligned}$$

Abbreviating $W_q = W(\psi_{q,-}, \psi_{q,+})$ we now compute

$$\frac{d}{dz}\alpha(z) = \frac{d}{dz}\Big(\frac{W}{W_q}\Big) = -\frac{\dot W_q}{W_q^2} W + \frac{1}{W_q}\Big(- \int_{-\infty}^{+\infty} (\psi_+\psi_- - \alpha \psi_{q,+}\psi_{q,-}) dr + \alpha \dot W_q\Big)$$
$$= -\frac{1}{W_q} \int_{-\infty}^{+\infty} \Big(\psi_+(z,r)\psi_-(z,r) - \alpha(z)\psi_{q,+}(z,r)\psi_{q,-}(z,r)\Big) dr,$$

which finishes the proof. □

Since $(H - z)^{-1} - (H_q - z)^{-1}$ is trace class with continuous integral kernel $G(z, x, x) - G_q(z, x, x)$, we have ([2])

$$\operatorname{tr}\big((H - z)^{-1} - (H_q - z)^{-1}\big) = \int_{-\infty}^{+\infty} (G(z, x, x) - G_q(z, x, x)) dx, \qquad z \in \mathbb{C}\backslash\sigma(H), \tag{4.3}$$

and the last result can be rephrased as

$$\frac{d}{dz} T(z) = T(z) \operatorname{tr}\big((H - z)^{-1} - (H_q - z)^{-1}\big), \qquad z \in \mathbb{C}\backslash\sigma(H). \tag{4.4}$$

As an immediate consequence we can establish the connection with Krein's spectral shift function ([22]). We refer to [38] for Krein's spectral shift theory in the case when only the resolvent difference is trace class; which is the case needed here.

Theorem 4.2. *The transmission coefficient $T(z)$ has the representation*

$$T(z) = \exp\Big(\int_{\mathbb{R}} \frac{\xi(\lambda) d\lambda}{\lambda - z}\Big), \tag{4.5}$$

where

$$\xi(\lambda) = \frac{1}{\pi} \lim_{\epsilon\downarrow 0} \arg T(\lambda + \mathrm{i}\epsilon) \tag{4.6}$$

is the spectral shift function of the pair H, H_q. Moreover, $(V - V_q)^{1/2}(H_q - z)^{-1}|V - V_q|^{1/2}$ is trace class and $T(z)$ is the perturbation determinant of the pair H and H_q:

$$T(z) = \det\big(\mathbb{1} + (V - V_q)^{1/2}(H_q - z)^{-1}|V - V_q|^{1/2}\big). \tag{4.7}$$

If in addition $(V - V_q)(H_q - z)^{-1}$ is trace class we have

$$T(z) = \det\big(\mathbb{1} + (V - V_q)(H_q - z)^{-1}\big). \tag{4.8}$$

Proof. The function $\operatorname{Im}\log(T(z))$ is a bounded harmonic function in the upper half-plane and hence has a Poisson representation (cf. [21])

$$\operatorname{Im}\log(T(z)) = \int_{\mathbb{R}} \frac{y}{(x - \lambda)^2 + y^2} \xi(\lambda) d\lambda. \quad z = x + \mathrm{i}y.$$

Moreover, by $\xi(\lambda) = 0$ for λ below the spectrum of H and $\xi(\lambda) = O(\lambda^{-1/2})$ as $\lambda \to +\infty$ (by Corollary 3.7) we obtain equality in (4.5) up to a real constant.

The missing constants follows since both sides tend to 1 as $z \to \infty$. Moreover, combining (4.5) with (4.4) we see

$$\mathrm{tr}\big((H-z)^{-1} - (H_q - z)^{-1}\big) = \int_{\mathbb{R}} \frac{\xi(\lambda) d\lambda}{(\lambda - z)^2},$$

which shows that $\xi(\lambda)$ is the spectral shift function.

That $T(z)$ is the perturbation determinant is standard if $(V - V_q)(H_q - z)^{-1}$ is trace class (see, e.g., [38]) for the slightly more general case when $(V-V_q)^{1/2}(H_q - z)^{-1}|V - V_q|^{1/2}$ is trace class we refer to [15, Sect. 4], [18, Sect. 7]. That this last condition holds will be shown in the next lemma below. □

To following result needed in the previous proof is of independent interest.

Lemma 4.3. *Assume* (3.2). *Then* $(V - V_q)^{1/2}(H_q - z)^{-1}|V - V_q|^{1/2}$ *is trace class. If we even have*

$$\|V - V_q\|_{2;1} = \sum_{n\in\mathbb{Z}} \left(\int_n^{n+1} |V(x) - V_q(x)|^2 \right)^{1/2} < \infty, \tag{4.9}$$

then $(V - V_q)(H_q - z)^{-1}$ *is trace class.*

Proof. To see the first claim we begin with the fact [32, Prop. 2.2] that $|V - V_q|^{1/2}(H_0 - z)^{-1}|V - V_q|^{1/2}$ is trace class, where $H_0 = -\frac{d^2}{dx^2}$. Let $z < 0$ and set $A(z) = (V - V_q)^{1/2}(H_0 - z)^{-1/2}$. Then $A(z)A(z)^* = |V - V_q|^{1/2}(H_0 - z)^{-1}|V - V_q|^{1/2}$ is trace class and thus $A(z)$ is Hilbert–Schmidt. In fact, since $A(z) = A(z_0)(H_0 - z_0)^{1/2}(H_0 - z)^{-1/2}$ this holds for all $z \in \rho(H_0)$ and not just for $z < 0$. Hence, using $(H_q - z)^{-1} = (H_0 - z)^{-1/2}C(z)(H_0 - z)^{-1/2}$, where $C(z)$ is bounded (cf. [37, Thm. 6.25]), we see $|V - V_q|^{1/2}(H_q - z)^{-1}|V - V_q|^{1/2} = AC(z)A^*$ which establishes the claim.

The see the second claim we again begin with the fact [33, Theorem 4.5] that (4.9) implies that $(V - V_q)(H_0 - z)^{-1}$ is trace class. Now the second resolvent equation $(H_q - z)^{-1} = (H_0 - z)^{-1} - (H_0 - z)^{-1}V_q(H_q - z)^{-1}$ establishes the claim since $V_q(H_q - z)^{-1}$ is bounded (cf. [37, Sect. 9.7]). □

Note that in the case $V_q = 0$ [33, Prop. 4.7] implies that the condition (4.9) is optimal. Moreover, the norm in (4.9) dominates the L^1 norm, $\|V\|_1 \le \|V\|_{2;1}$ by the Cauchy–Schwartz inequality, but the converse is of course not true (since (4.9) forces the function to be locally square integrable).

5. The transmission coefficient

Throughout this section we make the somewhat stronger assumption that

$$\int_{-\infty}^{+\infty} (1 + |x|)\big|V(x) - V_q(x)\big| dx < \infty \tag{5.1}$$

in order to ensure that there is only a finite number of eigenvalues in each gap [31]. Our aim is to reconstruct the transmission coefficient $T(z)$ from its boundary values and its poles. To this end, recall that $T(z)$ is meromorphic in $\mathbb{C}\backslash\sigma(H_q)$ with simple poles at the eigenvalues ρ_j of H. Moreover, for $z \in \sigma(H_q)$ the boundary values from the upper, respectively, lower, half-plane exist and satisfy $|T(z)|^2 = 1 - |R_\pm(z)|^2$, where $R_\pm(z)$ are the reflection coefficients defined in the previous section.

In the case where $V_q = 0$, this can be done via the classical Poisson–Jensen formula. In the more general setting here, the reconstruction needs to be done on the underlying Riemann surface. We essentially follow [36], where the analog problem for Jacobi operators was solved.

Denote by $\omega_{p\,q}$ the normalized Abelian differential of the third kind with poles at p and q. Then the Blaschke factor is defined by

$$B(p,\rho) = \exp\left(g(p,\rho)\right) = \exp\Big(\int_{E_0}^{p} \omega_{\rho\,\rho^*}\Big)\exp\Big(\int_{E(\rho)}^{\rho} \omega_{p\,p^*}\Big), \quad \pi(\rho)\in\mathbb{R}, \quad (5.2)$$

where $E(\rho)$ is E_0 if $\rho < E_0$ and either E_{2j-1} or E_{2j} if $\rho \in (E_{2j-1}, E_{2j})$, $1 \le j \le g$. It is a multi-valued function with a simple zero at ρ and simple pole at ρ^* satisfying $|B(p,\rho)| = 1$, $p \in \partial\Pi_+$. It is real-valued for $\pi(p) \in (-\infty, E_0)$ and satisfies

$$B(E_0,\rho) = 1 \quad \text{and} \quad B(p^*,\rho) = B(p,\rho^*) = B(p,\rho)^{-1}. \quad (5.3)$$

Then we have

Theorem 5.1. *The transmission coefficient is given by*

$$T(z,x) = \Big(\prod_{j=1}^{g} B(p,\rho_j)^{-1}\Big)\exp\Big(\frac{1}{2\pi\mathrm{i}}\int_{\partial\Pi_+}\log(1-|R_\pm|^2)\omega_{p\,p_\infty}\Big), \quad p=(z,+), \quad (5.4)$$

where we set $R_\pm(p) = R_\pm(z)$ for $p=(z,+)$ and $R_\pm(p) = \overline{R_\pm(z)}$ for $p=(z,-)$.

Proof. Just literally follow the argument in [36, Sect. 3]. □

Remark 5.2. A few remarks are in order:

(i) Using symmetry, $|R_\pm(p^*)| = |R_\pm(p)|$ for $p \in \partial\Pi_+$, of the integrand we can rewrite (5.4) as

$$T(p,x) = \Big(\prod_{j=1}^{g}\exp\Big(-\int_{E(\rho_j)}^{\rho_j}\omega_{p\,p^*}\Big)\Big)\exp\Big(\frac{1}{2\pi\mathrm{i}}\int_{\Sigma}\log(1-|R_\pm|^2)\omega_{p\,p^*}\Big), \quad (5.5)$$

where the integral over Σ is taken on the upper sheet.

(ii) There exist explicit formulas for Abelian differentials of the third kind:

$$\omega_{pq} = \left(\frac{R_{2g+1}^{1/2} + R_{2g+1}^{1/2}(p)}{2(\pi-\pi(p))} - \frac{R_{2g+1}^{1/2} + R_{2g+1}^{1/2}(q)}{2(\pi-\pi(q))} + P_{pq}(\pi)\right)\frac{d\pi}{R_{2g+1}^{1/2}},$$

$$\omega_{pp_\infty} = \left(-\frac{R_{2g+1}^{1/2} + R_{2g+1}^{1/2}(p)}{2(\pi-\pi(p))} + P_{pp_\infty}(\pi)\right)\frac{d\pi}{R_{2g+1}^{1/2}},$$

where $P_{pq}(z)$, $P_{pp_\infty}(z)$ are polynomials of degree $g-1$ which have to be determined from the normalization condition to have vanishing a-periods. In particular,

$$\omega_{pp^*} = \left(\frac{R_{2g+1}^{1/2}(p)}{\pi - \pi(p)} + P_{pp^*}(\pi)\right)\frac{d\pi}{R_{2g+1}^{1/2}}.$$

(iii) The function

$$T(p) = \begin{cases} T(z), & p=(z,+),\\ T(z)^{-1}, & p=(z,-),\end{cases}$$

solves the following scalar meromorphic Riemann–Hilbert factorization problem:

$$\begin{aligned} &T_+(p,x) = T_-(p,x)(1-|R(p)|^2), \quad p\in\partial\Pi_+,\\ &(T(p,x)) = \mathcal{D}_\rho^* - \mathcal{D}_\rho \\ &T(p_\infty,x)=1.\end{aligned} \tag{5.6}$$

Here the subscripts in $T_\pm(p)$ denote the limits from $\Pi_\pm$, respectively. Compare [23], [24].

As was pointed out in [36], this implies the following algebraic constraint on the scattering data.

Theorem 5.3. *The transmission coefficient T defined via* (5.4) *is single-valued if and only if the eigenvalues ρ_j and the reflection coefficients $R_\pm$ satisfy*

$$\sum_j \int_{\rho_j^*}^{\rho_j} \zeta_\ell - \frac{1}{2\pi\mathrm{i}}\int_{\partial\Pi_+}\log(1-|R_\pm|^2)\zeta_\ell \in \mathbb{Z}. \tag{5.7}$$

6. Conserved quantities of the KdV hierarchy

Finally we turn to solutions of the KdV hierarchy (see [13]). Let $V_q(x,t)$ be a finite-gap solution of some equation in the KdV hierarchy, $\mathrm{KdV}_r(V_q(x,t))=0$, and let $V(x,t)$ be another solution, $\mathrm{KdV}_r(V(x,t))=0$, such that $V(.,t)-V_q(.,t)$ is Schwartz class for all $t\in\mathbb{R}$. Existence of such solutions has been established only recently in [8].

Since the transmission coefficient $T(z,t)=T(z,0)\equiv T(z)$ is conserved (see [8] – formally this follows from unitary invariance of the determinant), conserved quantities of the KdV hierarchy can be obtained by computing the asymptotic expansion at ∞.

To this end, we begin by recalling the following well-known asymptotics for the Weyl m-functions in case of smooth potentials:

Lemma 6.1. *Suppose $V(x)\in C^\infty(\mathbb{R})$ is smooth. The Weyl m-functions have the following asymptotic expansion for large z*

$$m_\pm(z,x) \asymp \mathrm{i}\sqrt{z} \pm \sum_{n=1}^\infty \frac{\chi_n(x)}{(\pm 2\mathrm{i}\sqrt{z})^n}, \tag{6.1}$$

with coefficients defined recursively via

$$\chi_1(x) = V(x), \quad \chi_{n+1}(x) = -\frac{\partial}{\partial x}\chi_n(x) - \sum_{m=1}^{n-1} \chi_{n-m}(x)\chi_m(x). \qquad (6.2)$$

The corresponding expansion coefficients associated with V_q will be denoted by $\chi_{q,m}(x)$. It is also known that the even coefficients are complete differentials [4] and the first few are explicitly given by

$$\begin{aligned}
\chi_1(x) &= V(x),\\
\chi_2(x) &= -V'(x),\\
\chi_3(x) &= V''(x) - V(x)^2,\\
\chi_4(x) &= -V'''(x) + 4V(x)V'(x),\\
\chi_5(x) &= V''''(x) - 6V''(x)V(x) - 5V'(x)^2 + 2V(x)^3.
\end{aligned}$$

Theorem 6.2. *Suppose $V(x) - V_q(x) \in \mathcal{S}(\mathbb{R})$ is Schwartz. Then $\log T(z)$ has an asymptotic expansion around $z = \infty$:*

$$\log T(z) \asymp \mathrm{i}\sqrt{z} \sum_{k=1}^{\infty} \frac{\tau_k}{z^k}.$$

The quantities τ_k are given by

$$\tau_k = \int_{-\infty}^{\infty} \frac{\chi_{2k-1}(x) - \chi_{q,2k-1}(x)}{(-1)^k 2^{2k-1}} dx \qquad (6.3)$$

and are conserved quantities for the KdV hierarchy. Explicitly,

$$\begin{aligned}
\tau_1 &= -\frac{1}{2}\int_{-\infty}^{\infty} \big(V(x) - V_q(x)\big)dx,\\
\tau_2 &= -\frac{1}{8}\int_{-\infty}^{\infty} \big(V^2(x) - V_q^2(x)\big)dx,\\
\tau_3 &= -\frac{1}{32}\int_{-\infty}^{\infty} \big(2V^3(x) - 5V_x^2(x) - 6V_{xx}(x)V(x)\\
&\qquad - 2V_q^3(x) + 5V_{q,x}^2(x) + 6V_{q,xx}(x)V_q(x)\big)dx,
\end{aligned}$$

etc.

Proof. Represent the Jost solutions in the form

$$\psi_\pm(z,x) = \psi_{q,\pm}(z,x) \exp\left(\mp \int_x^{\pm\infty} \big(m_\pm(z,y) - m_{q,\pm}(z,y)\big)dy\right). \qquad (6.4)$$

Then iterating the Volterra integral equations (3.7) one sees that $\psi_\pm(z,x)$ have an asymptotic expansion uniformly with respect to x and given by

$$\log \frac{\psi_\pm(z,x)}{\psi_{q,\pm}(z,x)} \asymp -\sum_{n=1}^{\infty} \frac{1}{(\pm 2\mathrm{i}\sqrt{z})^n} \int_x^{\pm\infty} \big(\chi_n(y) - \chi_{q,n}(y)\big)dy. \qquad (6.5)$$

Then, letting $x \to \mp\infty$ using (3.12) yields (6.3). In particular, equality of the plus and minus cases shows that all even expansion coefficients must vanish (which alternatively also follows from the fact that the even expansion coefficients are complete differentials). □

Theorem 6.3. *Consider the expansion coefficients τ_k of $\log T(z)$ defined in (6.3). Then the following trace formulas are valid:*

$$\tau_k = 2\mathrm{i}\sum_{j=1}^{g}\int_{E(\rho_j)}^{\rho_j}\omega_{p_\infty,2k-2} - \frac{1}{\pi}\int_\Sigma \log|T|^2\omega_{p_\infty,2k-2}, \tag{6.6}$$

where $\omega_{p_\infty,k}$ is the Abelian differential of the second kind with a pole of order $k+2$ at p_∞.

Proof. From $\frac{d^k}{dz^k}\omega_{p\,E_0} = k!\omega_{p_\infty,k-1}$ we get that

$$\omega_{p\,E_0} = \omega_{p_\infty\,E_0} + \sum_{k=1}^{\infty}\zeta^k\omega_{p_\infty,k-1}, \qquad \zeta = z^{-1/2},$$

$$\omega_{p^*\,E_0} = \omega_{p_\infty\,E_0} + \sum_{k=1}^{\infty}\zeta^k\omega_{p_\infty,k-1}, \qquad \zeta = -z^{-1/2}.$$

Using this it follows

$$\omega_{p\,p^*} = \omega_{p\,E_0} - \omega_{p^*\,E_0} = 2\sum_{k=1}^{\infty}\omega_{p_\infty,2k-2}\zeta^{2k-1}, \quad \zeta = z^{-1/2}.$$

Hence we have

$$\begin{aligned}
&-\sum_{j=1}^{g}\int_{E(\rho_j)}^{\rho_j}\omega_{p\,p^*} + \frac{1}{2\pi\mathrm{i}}\int_\Sigma \log|T|^2\omega_{p\,p^*}\\
&= -\sum_{j=1}^{g}\int_{E(\rho_j)}^{\rho_j} 2\sum_{k=1}^{\infty}\zeta^{2k-1}\omega_{p_\infty,2k-2} + \frac{1}{\pi\mathrm{i}}\int_\Sigma \log|T|^2\sum_{k=1}^{\infty}\zeta^{2k-1}\omega_{p_\infty,2k-2}.
\end{aligned}$$

Thus, since $|T|^2 = 1 - |R_\pm|^2$ and $R_\pm(\lambda)$ decays faster than any polynomial as $\lambda \to \infty$ [8], one obtains

$$\log T(z) \asymp -\sum_{k=1}^{\infty}\left(2\sum_{j=1}^{\infty}\int_{E(\rho_j)}^{\rho_j}\omega_{p_\infty,2k-2} - \frac{1}{\pi\mathrm{i}}\int_\Sigma \log|T|^2\omega_{p_\infty,2k-2}\right)\zeta^{2k-1},$$

where $\zeta = z^{-1/2}$ denotes the local coordinate at $z = \infty$. □

Remark 6.4. The differentials $\omega_{p_\infty,2k-2}$, $k = 1, 2, \ldots$, are explicitly given by

$$\omega_{p_\infty,2k-2} = \left(\frac{\pi^{g+k-1}}{R_{2g+1}^{1/2}} + P_k(\pi)\right)d\pi. \tag{6.7}$$

Here $P_k(\pi)$ is a polynomial of degree $g + k - 2$ which has to be chosen such that all a-periods vanish.

Acknowledgment

G.T. would like to thank all organizers of the international conference on Operator Theory, Analysis and Mathematical Physics (OTAMP), Bedlewo, June 2008, for their kind invitation and the stimulating atmosphere during the meeting. We are indebted to Iryna Egorova and Fritz Gesztesy for several helpful discussions.

References

[1] A. Boutet de Monvel, I. Egorova, and G. Teschl, *Inverse scattering theory for one-dimensional Schrödinger operators with steplike finite-gap potentials*, J. d'Analyse Math. **106:1**, 271–316 (2008).

[2] C. Brislawn, *Kernels of trace class operators*, Proc. Amer. Math Soc. **104:4**, 1181–1190 (1988).

[3] P. Deift and R. Killip, *On the absolutely continuous spectrum of one-dimensional Schrödinger operators with square summable potentials*, Commun. Math. Phys. **203**, 341–347 (1999).

[4] P.G. Drazin and R.S. Johnson, *Solitons: An Introduction*, Cambridge Univ. Press, Cambridge, 1989.

[5] W. Eckhaus and A. Van Harten, *The Inverse Scattering Transformation and Solitons: An Introduction*, Math. Studies 50, North-Holland, Amsterdam, 1984.

[6] I. Egorova and G. Teschl, *On the Cauchy problem for the Korteweg–de Vries equation with steplike finite-gap initial data II. Perturbations with Finite Moments*, J. d'Analyse Math. (to appear).

[7] I. Egorova and G. Teschl, *A Paley-Wiener theorem for periodic scattering with applications to the Korteweg–de Vries equation*, Zh. Mat. Fiz. Anal. Geom. **6:1**, 21–33 (2010).

[8] I. Egorova, K. Grunert, and G. Teschl, *On the Cauchy problem for the Korteweg–de Vries equation with steplike finite-gap initial data I. Schwartz-type perturbations*, Nonlinearity **22**, 1431–1457 (2009).

[9] N.E. Firsova, *An inverse scattering problem for the perturbed Hill operator*, Mat. Zametki **18**, no. 6, 831–843 (1975).

[10] N.E. Firsova, *A direct and inverse scattering problem for a one-dimensional perturbed Hill operator* Matem. Sborn. (N.S.) **130(172)**, no. 3, 349–385 (1986).

[11] N.E. Firsova, *The Riemann surface of a quasimomentum, and scattering theory for a perturbed Hill operator* Mathematical questions in the theory of wave propagation, 7. Zap. Naučn. Sem. Leningrad. Otdel. Mat. Inst. Steklov (LOMI) **51**, 183–196 (1975).

[12] N.E. Firsova, *Solution of the Cauchy problem for the Korteweg-de Vries equation with initial data that are the sum of a periodic and a rapidly decreasing function*, Math. USSR-Sb. **63:1**, 257–265 (1989).

[13] F. Gesztesy and H. Holden, *Soliton Equations and Their Algebro-Geometric Solutions. Volume I:* (1 + 1)*-Dimensional Continuous Models.*, Cambridge Studies in Advanced Mathematics, Vol. 79, Cambridge University Press, Cambridge, 2003.

[14] F. Gesztesy and H. Holden, *Trace formulas and conservation laws for nonlinear evolution equations*, Rev. Math. Phys. **6**, 51–95 (1994).

[15] F. Gesztesy and K.A. Makarov, *(Modified) Fredholm determinants for operators with matrix-valued semi-separable integral kernels revisited*, Integral Eq. Operator Theory **47**, 457–497 (2003). (See also Erratum **48**, 425–426 (2004) and the corrected electronic only version in **48**, 561–602 (2004).)

[16] F. Gesztesy, A. Pushnitski, and B. Simon, *On the Koplienko spectral shift function. I. Basics.*, Zh. Mat. Fiz. Anal. Geom. **4:1**, 63–107 (2008).

[17] F. Gesztesy, R. Ratnaseelan, and G. Teschl, *The KdV hierarchy and associated trace formulas*, in "*Proceedings of the International Conference on Applications of Operator Theory*", (eds. I. Gohberg, P. Lancaster, and P.N. Shivakumar), Oper. Theory Adv. Appl., 87, Birkhäuser, Basel, 125–163 (1996).

[18] F. Gesztesy, Y. Latushkin, M. Mitrea, and M. Zinchenko, *Nonselfadjoint operators, infinite determinants, and some applications*, Russ. J. Math. Phys. **12:4**, 443–471 (2005).

[19] R. Jost and A. Pais, *On the scattering of a particle by a static potential*, Phys. Rev. **82**, 840–851 (1951).

[20] R. Killip and B. Simon, *Sum rules for Jacobi matrices and their applications to spectral theory*, Ann. of Math. (2) **158**, 253–321 (2003).

[21] P. Koosis, *Introduction to H_p Spaces*, 2nd ed., Cambridge Tracts in Mathematics **115**, Cambridge University Press, Cambridge, 1998.

[22] M.G. Krein, *Perturbation determinants and a formula for the traces of unitary and self-adjoint operators*, Soviet. Math. Dokl. **3**, 707–710 (1962).

[23] S. Kamvissis and G. Teschl, *Stability of periodic soliton equations under short range perturbations*, Phys. Lett. A **364:6**, 480–483 (2007).

[24] S. Kamvissis and G. Teschl, *Stability of the periodic Toda lattice under short range perturbations*, arXiv:0705.0346.

[25] H. Krüger and G. Teschl, *Relative oscillation theory, zeros of the Wronskian, and the spectral shift function*, Comm. Math. Phys. **87:2**, 613–640 (2009).

[26] H. Krüger and G. Teschl, *Effective Prüfer angles and relative oscillation criteria*, J. Diff. Eq. **245**, 3823–3848 (2008).

[27] A. Laptev, S. Naboko, and O. Safronov, *On new relations between spectral properties of Jacobi matrices and their coefficients*, Comm. Math. Phys. **241**, no. 1, 91–110 (2003).

[28] V.A. Marchenko, *Sturm-Liouville Operators and Applications*, Birkhäuser, Basel, 1986.

[29] J. Michor and G. Teschl, *Trace formulas for Jacobi operators in connection with scattering theory for quasi-periodic background*, Operator Theory, Analysis, and Mathematical Physics, J. Janas, et al. (eds.), 69–76, Oper. Theory Adv. Appl., 174, Birkhäuser, Basel, 2007.

[30] F. Nazarov, F. Peherstorfer, A. Volberg, and P. Yuditskii, *On generalized sum rules for Jacobi matrices*, Int. Math. Res. Not. **2005:3**, 155–186 (2005).

[31] F.S. Rofe-Beketov, *A finiteness test for the number of discrete levels which can be introduced into the gaps of the continuous spectrum by perturbations of a periodic potential*, Dokl. Akad. Nauk SSSR **156**, 515–518 (1964).

[32] B. Simon, *Resonances in one dimension and Fredholm determinants*, J. Funct. Anal. **178**, 396–420 (2000).

[33] B. Simon, *Trace Ideals and Their Applications*, 2nd ed., Amer. Math. Soc., Providence, 2005.

[34] B. Simon and A. Zlatoš, *Sum rules and the Szegö condition for orthogonal polynomials on the real line*, Comm. Math. Phys. **242:3**, 393–423 (2003).

[35] G. Teschl, *Jacobi Operators and Completely Integrable Nonlinear Lattices*, Math. Surv. and Mon. **72**, Amer. Math. Soc., Rhode Island, 2000.

[36] G. Teschl, *Algebro-geometric constraints on solitons with respect to quasi-periodic backgrounds*, Bull. London Math. Soc. **39:4**, 677–684 (2007).

[37] G. Teschl, *Mathematical Methods in Quantum Mechanics; With Applications to Schrödinger Operators*, Graduate Studies in Mathematics **99**, Amer. Math. Soc., Providence, 2009.

[38] D.R. Yafaev, *Mathematical Scattering Theory: General Theory*, Amer. Math. Soc., Rhode Island, 1992.

[39] A. Zlatoš, *Sum rules for Jacobi matrices and divergent Lieb-Thirring sums*, J. Funct. Anal. **225**, no. 2, 371–382 (2005).

Alice Mikikits-Leitner
Faculty of Mathematics
University of Vienna
Nordbergstrasse 15
A-1090 Wien, Austria
e-mail: `Alice.Mikikits-Leitner@univie.ac.at`
URL: `http://www.mat.univie.ac.at/~alice/`

Gerald Teschl
Faculty of Mathematics
University of Vienna
Nordbergstrasse 15
A-1090 Wien, Austria
and
International Erwin Schrödinger Institute for Mathematical Physics
Boltzmanngasse 9
A-1090 Wien, Austria
e-mail: `Gerald.Teschl@univie.ac.at`
URL: `http://www.mat.univie.ac.at/~gerald/`

Operator Theory:
Advances and Applications, Vol. 214, 125–134
© 2011 Springer Basel AG

Inner-outer Factorization for Weighted Schur Class Functions and Corresponding Invariant Subspaces

Alexey Tikhonov

Abstract. We prove the existence and uniqueness theorem of inner-outer factorization for weighted Schur class functions over multiply connected domains. Using our extension for the Sz.-Nagy-Foiaş functional model we give descriptions for absolutely continuous and singular subspaces and derive the inner-outer factorization theorem employing the link between factorizations and invariant subspaces.

Mathematics Subject Classification (2000). Primary 47A68; Secondary 47A56, 47A45.

Keywords. Factorization, invariant subspaces, functional model, Schur class.

0. Introduction

An operator-valued function $\Theta \in H^\infty(\mathbb{D}, \mathcal{L}(\mathfrak{N}_+, \mathfrak{N}_-))$ is called a function of *Schur class* if $||\Theta||_\infty \leq 1$, that is,

$$\forall\, |z| < 1 \ \ \forall\, n \in \mathfrak{N}_+ \quad ||\Theta(z)n|| \leq ||n||\,,$$

where $\mathbb{D}$ is the open unit disk, $\mathfrak{N}_\pm$ are separable Hilbert spaces, H^∞ is the Hardy space [1, 2]. Schur class functions play a remarkable role in many mathematical disciplines, in function theory [1, 3, 4], in operator theory [2, 4], in mathematical physics (e.g., scattering theory) [5] and so on. Theory of such functions is well developed and a pertinent survey can be found, e.g., in [2] or [4].

The inner-outer factorization, i.e., a representation of an analytic operator function $\Theta(z)$ of Schur class on the unit disk in the form $\Theta(z) = \Theta_i(z)\Theta_e(z)$, where the boundary values for the inner function $\Theta_i(z)$ are isometrical almost everywhere on the unit circle and the outer function $\Theta_e(z)$ has no inner divisors,

Acknowledgement. Research for this article was supported by INTAS grant, project 05-1000008-7883.

is well known [1, 2]. This factorization is fundamental in function theory, operator theory, and for prediction theory of stochastic processes.

In this paper we prove the existence and uniqueness theorem of inner-outer factorization for *weighted Schur class* functions over multiply connected domains

$$S_\Xi := \{ (\Theta, \Xi_+, \Xi_-) : \Theta \in H^\infty(G_+, \mathcal{L}(\mathfrak{N}_+, \mathfrak{N}_-)), \\ \forall \zeta \in C \ \forall n \in \mathfrak{N}_+ \quad ||\Theta(\zeta)n||_{-,\zeta} \overset{a.e.}{\leq} ||n||_{+,\zeta} \},$$

where G_+ is a finitely connected domain of the complex plane bounded by a rectifiable Carleson curve C, $G_- = \mathbb{C} \setminus \operatorname{clos} G_+$ and $\infty \in G_-$; $\Theta(\zeta)$ are boundary values of $\Theta(z)$, $z \in G_+$; $\Xi_\pm$ are operator-valued weights such that $\Xi_\pm, \Xi_\pm^{-1} \in L^\infty(C, \mathcal{L}(\mathfrak{N}_\pm))$, $\Xi_\pm(\zeta) \geq 0$, $\zeta \in C$, and $||n||_{\pm,\zeta} := (\Xi_\pm(\zeta)n, n)^{1/2}$, $n \in \mathfrak{N}_\pm$.

Note that we consider the triplets $\theta = (\Theta, \Xi_+, \Xi_-)$ of the function Θ and the two weights $\Xi_\pm$, but we will use for them both the terms "triplet" and "weighted Schur class function" interchangeably.

Definitions. A triplet $\theta = (\Theta, \Xi_+, \Xi_-) \in S_\Xi$ is called

- Ξ-*inner* if $||\Theta(\zeta)n||_{-,\zeta} \overset{a.e.}{=} ||n||_{+,\zeta}$;
- Ξ-*outer* if it has no non-trivial left Ξ-inner divisors;
- Ξ-*unitary constant* if θ is Ξ-inner and $\Theta^{-1} \in H^\infty(G_+, \mathcal{L}(\mathfrak{N}_-, \mathfrak{N}_+))$.

We define the product

$$\theta_2\theta_1 := (\Theta_2\Theta_1, \Xi_{1+}, \Xi_{2-})$$

of two triplets $\theta_1 = (\Theta_1, \Xi_{1+}, \Xi_{1-})$ and $\theta_2 = (\Theta_2, \Xi_{2+}, \Xi_{2-})$ provided $\Xi_{2+} = \Xi_{1-}$. We are going to prove the following

Main Theorem. *Let* $\theta \in S_\Xi$. *Then* $\exists\, \theta_{\text{in}}, \theta_{\text{out}} \in S_\Xi$ *such that*

$$\theta = \theta_{\text{in}}\theta_{\text{out}},$$

where θ_{in} *is* Ξ*-inner and* θ_{out} *is* Ξ*-outer triplets. This factorization is unique (up to* Ξ*-unitary constant factors).*

The extension of Sz.-Nagy-Foiaş's functional model from [6, 7] (see some details in the following section) is the main tool for the proof. In outline, we follow the way of proof from [8].

Besides, we describe certain subspaces (singular and absolutely continuous invariant subspaces) in the functional model, which correspond to the factors in the inner-outer factorization, and give another characterization for Ξ-outer functions.

1. Factorizations and invariant subspaces

Here we survey the connection mentioned in the title of section. This fundamental link between factorizations of characteristic functions and invariant subspaces dates back to [9, 10]. In [11] we extended results developed in [2] to the class of weighted Schur functions and the paper [11] is a basic reference for presented information.

First, we recall the construction of free *functional model* of Sz.-Nagy-Foiaş type. Let $\Pi = (\pi_+, \pi_-)$ be a pair of operators $\pi_\pm \in \mathcal{L}(L^2(C, \mathfrak{N}_\pm), \mathcal{H})$ such that

(i)$_1$ $(\pi_\pm^* \pi_\pm) z = z(\pi_\pm^* \pi_\pm)$; (i)$_2$ $\pi_\pm^* \pi_\pm \gg 0$;

(ii)$_1$ $(\pi_-^\dagger \pi_+) z = z(\pi_-^\dagger \pi_+)$; (ii)$_2$ $P_-(\pi_-^\dagger \pi_+) P_+ = 0$;

(iii) $\operatorname{Ran} \pi_+ \vee \operatorname{Ran} \pi_- = \mathcal{H}$,

where $\mathfrak{N}_\pm, \mathcal{H}$ are separable Hilbert spaces; $A \gg 0$ means that $\exists\, c > 0$ such that $\forall u\ (Au, u) \geq c(u, u)$; the (nonorthogonal) projections $P_\pm$ are uniquely determined by conditions $\operatorname{Ran} P_\pm = E^2(G_\pm, \mathfrak{N}_\pm)$ and $\operatorname{Ker} P_\pm = E^2(G_\mp, \mathfrak{N}_\pm)$ (since the curve C is a Carleson curve, the projections $P_\pm$ are bounded); the spaces $E^2(G_\pm, \mathfrak{N}_\pm)$ are Smirnov's spaces [12] of vector-valued functions with values in $\mathfrak{N}_\pm$; the operators $\pi_\pm^\dagger$ are adjoint to $\pi_\pm$ if we regard $\pi_\pm \colon L^2(C, \Xi_\pm) \to \mathcal{H}$ as operators acting from weighted L^2 spaces with operator-valued weights $\Xi_\pm = \pi_\pm^* \pi_\pm$. In this interpretation $\pi_\pm$ are isometries.

There is a one-to-one correspondence between weighted Schur class functions and functional models. The mapping $\Pi \mapsto \theta$ defined by the formula

$$\theta = (\pi_-^\dagger \pi_+, \pi_+^* \pi_+, \pi_-^* \pi_-)$$

is one of the directions of this correspondence. The operator $\pi_-^\dagger \pi_+$ must be regarded as an analytic continuation of operator of multiplication by operator-valued function on the curve C into the domain G_+. Conversely, for a given $\theta \in S_\Xi$, it is possible to construct (up to unitary equivalence) a functional model Π such that $\theta = (\pi_-^\dagger \pi_+, \pi_+^* \pi_+, \pi_-^* \pi_-)$.

Further, there exists a normal operator $\mathcal{U} \in \mathcal{L}(\mathcal{H})$ with absolutely continuous spectrum on C, which is uniquely determined by conditions $\mathcal{U}\pi_\pm = \pi_\pm z$. Define also the projection $P_\theta := (I - \pi_+ P_+ \pi_+^\dagger)(I - \pi_- P_- \pi_-^\dagger)$ onto the subspace $\mathcal{K}_\theta := \operatorname{Ran} P_\theta$. Then one can consider the "main" operator $T \in \mathcal{L}(\mathcal{K}_\theta)$

$$Tf := \mathcal{U} f - \pi_+ \frac{1}{2\pi i} \int_C (\pi_+^\dagger f)(z)\, dz\,, \quad f \in \mathcal{K}_\theta\,.$$

Now we are ready to state our extension of the above-mentioned fundamental link between factorizations of characteristic function $\Theta(z)$ and invariant subspaces of operator T.

Fix weighted Schur class function $\theta = (\Theta, \Xi_+, \Xi_-)$ and the corresponding functional model $\Pi = (\pi_+, \pi_-)$.

Let $L \subset \mathcal{K}_\theta$ be an invariant subspace, $(T - z)^{-1} L \subset L$ for all $z \in G_-$. Then there exists an operator $\pi \in \mathcal{L}(L^2(C, \mathfrak{N}), \mathcal{H})$ such that

$$\begin{array}{ll} (\pi^* \pi) z = z(\pi^* \pi); & \pi^* \pi \gg 0; \\ (\pi^\dagger \pi_+) z = z(\pi^\dagger \pi_+); & (\pi_-^\dagger \pi) z = z(\pi_-^\dagger \pi); \\ P_-(\pi^\dagger \pi_+) P_+ = 0; & P_-(\pi_-^\dagger \pi) P_+ = 0; \\ \pi_-^\dagger \pi_+ = (\pi_-^\dagger \pi)(\pi^\dagger \pi_+) & \end{array} \tag{F\&I}$$

and $L = \mathcal{K}_{\theta_1} = \operatorname{Ran} P_{\theta_1}$, where $P_{\theta_1} = P_{\pi_+ \vee \pi}(I - \pi_+ P_+ \pi_+^\dagger)(I - \pi P_- \pi^\dagger)$ and $P_{\pi_+ \vee \pi}$ is the orthoprojetion onto $\operatorname{Ran} \pi_+ \vee \operatorname{Ran} \pi$. Thus, if we denote $\Theta_1 = \pi^\dagger \pi_+$,

$\Theta_2 = \pi_-^\dagger\pi$, $\Xi = \pi^*\pi$, $\theta_1 = (\Theta_1, \Xi_+, \Xi)$, $\theta_2 = (\Theta_2, \Xi, \Xi_-)$, then we obtain the regular factorization $\theta = \theta_2\theta_1$. Recall [2] that a factorization $\theta = \theta_2\theta_1$ is called regular if

$$\operatorname{Ran}(I - \Theta_2^\dagger(z)\Theta_2(z))^{1/2} \cap \operatorname{Ran}(I - \Theta_1(z)\Theta_1^\dagger(z))^{1/2} = \{0\}, \quad \text{a.e. } z \in \mathbb{T},$$

where $\Theta_1^\dagger$ and $\Theta_2^\dagger$ are adjoint to the operators Θ_1 and Θ_2 in the corresponding weighted L^2-spaces, respectively.

Conversely, if a factorization $\theta = \theta_2\theta_1$ is regular, then there exists an operator $\pi \in \mathcal{L}(L^2(C, \mathfrak{N}), \mathcal{H})$ satisfying the conditions (F&I) such that $\Theta_1 = \pi^\dagger\pi_+$ and $\Theta_2 = \pi_-^\dagger\pi$. Then the subspace $\mathcal{K}_{\theta_1}$ is invariant under the resolvent $(T - z)^{-1}$, $z \in G_-$.

The above-described correspondence between factorizations and invariant subspaces is one-to-one (up to Ξ-unitary constant factors). Moreover, this correspondence is order-preserving. For factorizations $\theta = \theta_2'\theta_1'$, $\theta = \theta_2\theta_1$ and for corresponding invariant subspaces $\mathcal{K}_{\theta_1'}$, $\mathcal{K}_{\theta_1}$, we have

$$\mathcal{K}_{\theta_1'} \subset \mathcal{K}_{\theta_1} \quad \Longleftrightarrow \quad \theta_2'\theta_1' \prec \theta_2\theta_1 .$$

We say that $\theta_2'\theta_1' \prec \theta_2\theta_1$ if there exists a weighted Schur class function ϑ such that $\theta_1 = \vartheta\theta_1'$ and $\theta_2' = \theta_2\vartheta$, where all the factorizations are regular.

Additionally we will need to know more details about geometry of invariant subspaces. Let $\theta = \theta_2\theta_1$ be a regular factorization, (π_+, π_-) be a functional model corresponding to θ and π be an operator corresponding to the factorization. We saw $\mathcal{K}_{\theta_1} \subset \mathcal{K}_\theta$. But, in general, $\mathcal{K}_{\theta_2} \not\subset \mathcal{K}_\theta$. Here $\mathcal{K}_{\theta_2} = \operatorname{Ran} P_{\theta_2}$ and $P_{\theta_2} = P_{\pi_- \vee \pi}(I - \pi P_+ \pi^\dagger)(I - \pi_- P_- \pi_-^\dagger)$. Nevertheless, we have

$$\begin{aligned} &P_{\theta_1}P_{\theta_2} = P_{\theta_2}P_{\theta_1} = 0, \quad P_\theta P_{\theta_1} = P_{\theta_1}, \quad P_{\theta_2}P_\theta = P_{\theta_2}, \\ &P_\theta(P_{\theta_1} + P_{\theta_2})P_\theta = P_\theta, \quad (P_{\theta_1} + P_{\theta_2})P_\theta(P_{\theta_1} + P_{\theta_2}) = (P_{\theta_1} + P_{\theta_2}). \end{aligned}$$

2. Inner-outer factorization

First, we note that for a weighted Schur class function $\theta = (\Theta, \Xi_+, \Xi_-) \in S_\Xi$ one can consider the dual triplet $\theta_* := (\Theta^\sim, \Xi_-^{\sim -1}, \Xi_+^{\sim -1}) \in S_\Xi$, where $A^\sim(z) := A(\bar{z})^*$. In this connection, we define the dual notions.

Definitions. A triplet $\theta = (\Theta, \Xi_+, \Xi_-) \in S_\Xi$ is called

- $*$-Ξ-*inner* if θ_* is Ξ-*inner*;
- $*$-Ξ-*outer* if θ_* is Ξ-*outer*.

Thus it suffices to establish the existence and uniqueness theorem for the dual $*$-outer-inner factorization $\theta_{*\,\text{out}}\theta_{*\,\text{in}}$, which is connected with inner-outer factorization $\theta_{\text{in}}\theta_{\text{out}}$ by the identity $(\theta_{\text{in}}\theta_{\text{out}})_* = \theta_{*\,\text{out}}\theta_{*\,\text{in}}$.

We divide the proof of the Main Theorem into parts, which we arrange as separate assertions. From now on and to the end of the section, we fix weighted Schur class function $\theta = (\Theta, \Xi_+, \Xi_-)$ and the corresponding functional model $\Pi = (\pi_+, \pi_-)$.

We start with

Lemma 2.1. *Let* $\mathcal{K}_{*\,\mathrm{in}} := \{f \in \mathcal{K}_\theta : (I - \pi_+\pi_+^\dagger)f = 0\}$. *Then* $\mathcal{K}_{*\,\mathrm{in}}$ *is an invariant subspace* $(T-z)^{-1}\mathcal{K}_{*\,\mathrm{in}} \subset \mathcal{K}_{*\,\mathrm{in}}\,, z \in G_-$. *If* $\pi \in \mathcal{L}(L^2(C,\mathfrak{N}),\mathcal{H})$ *corresponds to the invariant subspace* $\mathcal{K}_{*\,\mathrm{in}}$ *and* $\Theta_{*\,\mathrm{in}} := \pi^\dagger\pi_+$, *then* $\Theta_{*\,\mathrm{in}}\Theta_{*\,\mathrm{in}}^\dagger = I$.

Proof. For $f \in \mathcal{K}_{*\,\mathrm{in}}$ and $z \in G_-$, we have

$$(I - \pi_+\pi_+^\dagger)(T-z)^{-1}f = (I - \pi_+\pi_+^\dagger)(\mathcal{U}-z)^{-1}(f - \pi_+(\pi_+^\dagger f)(z))$$
$$= (\mathcal{U}-z)^{-1}((I - \pi_+\pi_+^\dagger)f - (I - \pi_+\pi_+^\dagger)\pi_+(\pi_+^\dagger f)(z)) = (\mathcal{U}-z)^{-1}0 = 0\,.$$

Hence $(T-z)^{-1}f \in \mathcal{K}_{*\,\mathrm{in}}$, that is, the subspace $\mathcal{K}_{*\,\mathrm{in}}$ is invariant. Further,

$$0 = (I - \pi_+\pi_+^\dagger)P_{\theta_1}\pi = (I - \pi_+\pi_+^\dagger)(I - \pi_+P_+\pi_+^\dagger)(I - \pi P_-\pi^\dagger)\pi$$
$$= (I - \pi_+\pi_+^\dagger)(\pi - \pi P_-) = (I - \pi_+\pi_+^\dagger)\pi P_+$$

and we get $(I - \pi_+\pi_+^\dagger)\pi = 0$. Then

$$I - \Theta_{*\,\mathrm{in}}\Theta_{*\,\mathrm{in}}^\dagger = I - \pi^\dagger\pi_+\pi_+^\dagger\pi = \pi^\dagger\pi - \pi^\dagger\pi_+\pi_+^\dagger\pi = \pi^\dagger(I - \pi_+\pi_+^\dagger)\pi = 0\,. \qquad \square$$

We have obtained the factorization $\theta = \theta_{*\,\mathrm{out}}\theta_{*\,\mathrm{in}}$, which corresponds to the subspace $\mathcal{K}_{*\,\mathrm{in}}$. The factor $\theta_{*\,\mathrm{in}}$ is $*$-Ξ-inner and therefore the factorization is regular. This factorization is extremal as it is shown in

Lemma 2.2. *Let* $\theta = \theta_2'\theta_1'$ *and* θ_1' *is* $*$-Ξ-*inner. Then* $\theta_2'\theta_1' \prec \theta_{*\,\mathrm{out}}\theta_{*\,\mathrm{in}}$.

Proof. Since θ_1' is $*$-Ξ-inner, the factorization $\theta = \theta_2'\theta_1'$ is regular. Then it suffices to check that $\mathcal{K}_{\theta_1'} \subset \mathcal{K}_{*\,\mathrm{in}}$. Let an operator $\pi' \in \mathcal{L}(L^2(C,\mathfrak{N}'),\mathcal{H})$ correspond to the factorization $\theta = \theta_2'\theta_1'$. We have

$$\pi'^\dagger(I - \pi_+\pi_+^\dagger)(I - \pi_+\pi_+^\dagger)\pi' = \pi'^\dagger(I - \pi_+\pi_+^\dagger)\pi' = I - \theta_1'\theta_1'^\dagger = 0\,.$$

Hence $(I - \pi_+\pi_+^\dagger)\pi' = 0$ and $(I - \pi_+\pi_+^\dagger)P_{\pi_+\vee\pi'} = 0$. Then $\forall f \in \mathcal{K}_{\theta_1'} \subset \mathcal{K}_\theta$

$$(I - \pi_+\pi_+^\dagger)f = (I - \pi_+\pi_+^\dagger)P_{\pi_+\vee\pi'}(I - \pi_+P_+\pi_+^\dagger)(I - \pi'P_-\pi'^\dagger)f = 0$$

and therefore $\mathcal{K}_{\theta_1'} \subset \mathcal{K}_{*\,\mathrm{in}}$. $\square$

Corollary. *The factor* $\theta_{*\,\mathrm{out}}$ *is* $*$-Ξ-*outer.*

Proof. Let $\theta_{*\,\mathrm{out}} = \theta_2\theta_1$ and θ_1 be $*$-Ξ-inner. Since $\theta_1\theta_{*\,\mathrm{in}}$ is $*$-Ξ-inner and $\theta = \theta_{*\,\mathrm{out}}\theta_{*\,\mathrm{in}} = \theta_2(\theta_1\theta_{*\,\mathrm{in}})$, we get $\theta_2(\theta_1\theta_{*\,\mathrm{in}}) \prec \theta_{*\,\mathrm{out}}\theta_{*\,\mathrm{in}}$. Then there exists $\vartheta \in S_\Xi$ such that $\theta_{*\,\mathrm{in}} = \vartheta\theta_1\theta_{*\,\mathrm{in}}$. Taking the main component $\Theta_{*\,\mathrm{in}}$ of the triplet $\theta_{*\,\mathrm{in}}$, we can rewrite the condition that $\theta_{*\,\mathrm{in}}$ $*$-Ξ-inner in the form $\Theta_{*\,\mathrm{in}}\Theta_{*\,\mathrm{in}}^\dagger = I$. Similarly, $I = \Theta_1\Theta_1^\dagger$. Multiplying the identity $\Theta_{*\,\mathrm{in}} = \varTheta\Theta_1\Theta_{*\,\mathrm{in}}$ by $\Theta_{*\,\mathrm{in}}^\dagger$, we get $I = \varTheta\Theta_1$. Together with $I = \Theta_1\Theta_1^\dagger$ it implies $\Theta_1^{-1} = \varTheta \in H^\infty(G_+)$, that is, θ_1 is a Ξ-unitary constant. $\square$

Lemma 2.3. *Let* $\theta_{*\,\mathrm{out}}'$ *be* $*$-Ξ-*outer,* $\theta_{*\,\mathrm{in}}'$ *be* $*$-Ξ-*inner and* $\theta = \theta_{*\,\mathrm{out}}'\theta_{*\,\mathrm{in}}'$. *Then there exists a* Ξ-*unitary constant* ϑ *such that* $\theta_{*\,\mathrm{out}}' = \theta_{*\,\mathrm{out}}\vartheta$ *and* $\theta_{*\,\mathrm{in}} = \vartheta\theta_{*\,\mathrm{in}}'$.

Proof. By Lemma 2.2, we have $\theta'_{*\,\mathrm{out}}\theta'_{*\,\mathrm{in}} \prec \theta_{*\,\mathrm{out}}\theta_{*\,\mathrm{in}}$. Hence there exists $\vartheta \in S_\Xi$ such that $\theta'_{*\,\mathrm{out}} = \theta_{*\,\mathrm{out}}\vartheta$ and $\theta_{*\,\mathrm{in}} = \vartheta\theta'_{*\,\mathrm{in}}$. Using the elementary lemma (*Let $A = BC$ and A, B be isometries $\Rightarrow$ C is an isometry*), we get ϑ is $*$-Ξ-inner. Since $\theta'_{*\,\mathrm{out}}$ is $*$-Ξ-outer, the triplet ϑ is a Ξ-unitary constant. □

Thus we have obtained the existence and uniqueness theorem of $*$-inner-outer factorization

Theorem 2.4. *Let $\theta \in S_\Xi$. Then $\exists\, \theta_{*\,\mathrm{in}}, \theta_{*\,\mathrm{out}} \in S_\Xi$ such that*

$$\theta = \theta_{*\,\mathrm{out}}\theta_{*\,\mathrm{in}}\,,$$

where $\theta_{*\,\mathrm{in}}$ is $*$-Ξ-inner and $\theta_{*\,\mathrm{out}}$ is $*$-Ξ-outer triplets. This factorization is unique (up to Ξ-unitary constant factors).

The Main Theorem is a direct consequence of this theorem.

Remark. For simply connected domains, using factorizations $\Xi_\pm = \chi^*_\pm\chi_\pm$, it is possible to reduce this theorem to the classical inner-outer factorization theorem [2]. But in the multiply connected case, the factors $\chi_\pm$ can be multi-valued. This multi-valuedness requires uniformization technique or analytic vector bundles [13] (see [14] with outline of possible proof of the theorem). However we prefer to work with the domain G_+ directly (than with its universal cover), especially because it is important for us to expose the link of inner-outer factorization with the invariant subspaces $\mathcal{K}_{*\,\mathrm{in}}$ and $\mathcal{K}_{\mathrm{out}}$. We have already described the subspace $\mathcal{K}_{*\,\mathrm{in}}$ and are going to obtain description for $\mathcal{K}_{\mathrm{out}}$ in the last section of the paper.

Example. At the end of the section we present an example of non-trivial Ξ-unitary constant $\theta = (\Theta, \Xi_+, \Xi_-)$. First, we consider the character-automorphic function

$$f_{k,r,R}(z) = \left(\frac{z}{r}\right)^{\frac{\ln k}{\ln R - \ln r}}, \quad r < |z| < R$$

with the properties

$$f_{k,r,R}(ze^{it})|_{t=2\pi} = e^{\frac{2\pi i \ln k}{\ln R - \ln r}} f_{k,r,R}(z)$$

$$|f_{k,r,R}(re^{it})| = 1\,, \quad |f_{k,r,R}(Re^{it})| = k\,.$$

This function is unique (up to unimodular constant multiplier) among character-automorphic functions such that $|f(re^{it})|=1$,$|f(Re^{it})| = k$ and $0<c<|f(z)|<C$.

We put $\Theta(z) = f_{k,r,R}(z) = \frac{z}{r}$, with $\ln k = \ln R - \ln r$ and

$$\Xi_+(\zeta) = \begin{cases} 1 & , |\zeta| = r \\ k^2k_1^2 & , |\zeta| = R \end{cases}, \quad \Xi_-(\zeta) = \begin{cases} 1 & , |\zeta| = r \\ k_1^2 & , |\zeta| = R \end{cases}.$$

The function $\Theta(z)$ is single-valued and the triplet $\theta = (\Theta, \Xi_+, \Xi_-)$ is a Ξ-unitary constant. But, if $\frac{\ln k_1}{\ln R - \ln r} \notin \mathbb{Z}$, the corresponding functions $\chi_+(z) = f_{k_1k,r,R}(z)$ and $\chi_-(z) = f_{k_1,r,R}(z)$ are multi-valued and it is impossible (using only single-valued functions) to reduce $\Theta(z)$ to a unitary (in the usual meaning) constant.

3. Description of outer functions and corresponding invariant subspaces

For $\theta \in S_\Xi$, we define $E_\theta := \{u \in E^2(G_-) : \Theta u \in E^2(G_+)\}$, (see, e.g., [7]).

Lemma 3.1. $\mathcal{K}_{*\,\mathrm{in}} = \{\pi_+ u : u \in E_\theta\}$.

Proof. Let $f \in \mathcal{K}_{*\,\mathrm{in}}$. Then $(I - \pi_+\pi_+^\dagger)f = 0$ and therefore $f = \pi_+ u$, where $u = \pi_+^\dagger f$. Since $f \in \mathcal{K}_\theta$, we have $u = \pi_+^\dagger f \in E^2(G_-)$ and $\Theta u = \pi_-^\dagger \pi_+ \pi_+^\dagger f = \pi_-^\dagger f \in E^2(G_+)$. Hence, $f \in \{\pi_+ u : u \in E_\theta\}$.

Conversely, let $f \in \{\pi_+ u : u \in E_\theta\}$. Then $(I - \pi_+\pi_+^\dagger)f = (I - \pi_+\pi_+^\dagger)\pi_+ u = \pi_+(I - \pi_+^\dagger \pi_+)u = \pi_+(I - I)u = 0$, $\pi_+^\dagger f = u \in E^2(G_-)$, and $\pi_-^\dagger f = \Theta u \in E^2(G_+)$. Hence, $f \in \mathcal{K}_{*\,\mathrm{in}}$. □

We need to make use of the following pairing

$$\langle u, v\rangle_C := \frac{1}{2\pi i}\int_C (u(z), v(\bar{z}))_{\mathfrak{N}}\, dz\ ,\ u \in L^2(C, \mathfrak{N}),\ v \in L^2(\bar{C}, \mathfrak{N})\,.$$

It can be shown in the usual way that any linear continuous functional $F \in E^2(\bar{G}_\pm)^*$ can be represented in the form $F(u) = \langle u, v\rangle$ with $v \in E^2(G_\mp)$. Thus, $E^2(\bar{G}_\pm)^* \simeq E^2(G_\mp)$ and it can easily be checked that $E^2(G_\pm)^{\langle\perp\rangle} = E^2(\bar{G}_\pm)$.

Theorem 3.2. $E_\theta = \{0\} \iff \operatorname{clos}\Theta^\sim E^2(\bar{G}_+) = E^2(\bar{G}_+)\,.$

Proof. Let $u \in E_\theta$. Then $\forall v \in E^2(\bar{G}_+)\quad 0 = \langle \Theta u, v\rangle = \langle u, \Theta^\sim v\rangle$ and therefore $u\ \langle\perp\rangle\ \operatorname{clos}\Theta^\sim E^2(\bar{G}_+)$. If $\operatorname{clos}\Theta^\sim E^2(\bar{G}_+) = E^2(\bar{G}_+)$, then $u \in E^2(G_+)$. Since $u \in E^2(G_-)$, we have $u = 0$ and $E_\theta = \{0\}$.

Conversely, let $E_\theta = \{0\}$. Assume that $\exists\, w \in E^2(\bar{G}_+) \setminus \operatorname{clos}\Theta^\sim E^2(\bar{G}_+)$ such that $w \neq 0$. Then, by Hahn-Banach theorem, there exists $F \in E^2(\bar{G}_\pm)^*$ such that $F(\operatorname{clos}\Theta^\sim E^2(\bar{G}_+)) = 0$ and $F(w) \neq 0$. Since $E^2(\bar{G}_\pm)^* \simeq E^2(G_\mp)$, there exists $v \in E^2(G_-)$ such that $0 = \left\langle \Theta^\sim E^2(\bar{G}_+), v\right\rangle = \left\langle E^2(\bar{G}_+), \Theta v\right\rangle$ and therefore $\Theta v \in E^2(G_+)$. Hence, $v \in E_\theta$ and $v = 0$. This contradicts $\langle w, v\rangle \neq 0$. Thus, $\operatorname{clos}\Theta^\sim E^2(\bar{G}_+) = E^2(\bar{G}_+)$. □

Corollary. *$\theta \in S_\Xi$ is $*$-Ξ-outer if and only if* $\operatorname{clos}\Theta^\sim E^2(\bar{G}_+) = E^2(\bar{G}_+)\,.$

Proof. By Lemma 3.1, $\mathcal{K}_{*\,\mathrm{in}} = \{0\} \Leftrightarrow E_\theta = \{0\}$. □

Thus we arrive at the following description for Ξ-outer functions

Theorem 3.3. *$\theta \in S_\Xi$ is Ξ-outer if and only if* $\operatorname{clos}\Theta E^2(G_+) = E^2(G_+)\,.$

To describe the corresponding invariant subspace $\mathcal{K}_{\mathrm{out}}$ we need to prove the following two lemmas.

Lemma 3.4. *Let P, P_1, P_2 be projections in $\mathcal{H}$ such that*

$$\begin{aligned} &P_1P_2 = P_2P_1 = 0, \quad PP_1 = P_1, \quad P_2P = P_2,\\ &P(P_1 + P_2)P = P, \quad (P_1 + P_2)P(P_1 + P_2) = (P_1 + P_2). \end{aligned}$$

Let $\mathcal{K}_* = \operatorname{Ran} P^*$, $H_1 = \operatorname{Ran} P_1$, $H_{2*} = \operatorname{Ran} P_2^*$. Then $H_{2*} = \mathcal{K}_* \bigcap H_1^\perp$.

Proof. Let $g \in \mathcal{K}_* \bigcap H_1^\perp$. Then $\forall f \in \mathcal{H} \quad 0 = (P_1 f, g) = (f, P_1^* g)$ and therefore $P_1^* g = 0$. Further,

$$P_2^* g = P^* P_2^* g = P^*(P_1^* + P_2^*) g = P^*(P_1^* + P_2^*) P^* g = P^* g = g\,.$$

Hence $g \in H_{2*}$ and $\mathcal{K}_* \bigcap H_1^\perp \subset H_{2*}$.

Conversely, let $g \in H_{2*}$. Then $\forall f \in \mathcal{H}$

$$(P_1 f, g) = (P_1 f, P_2^* g) = (P_2 P_1 f, g) = 0\,.$$

Hence $g \perp H_1$. Since $P^* P_2^* = P_2^*$, we have $g \in \mathcal{K}_*$ and $H_{2*} \subset \mathcal{K}_* \bigcap H_1^\perp$. □

Lemma 3.5. *Let P be projection in $\mathcal{H}$, $\mathcal{K} = \operatorname{Ran} P$, $\mathcal{K}_* = \operatorname{Ran} P^*$. Let M be a linear closed subspace in $\mathcal{H}$. Then $\mathcal{K}_* \bigcap (\mathcal{K} \bigcap M)^\perp = \operatorname{clos} P^* M^\perp$.*

Proof. Let $f = P^* g$, $g \perp M$. Obviously, $f \in \mathcal{K}_*$ and $\forall h \in \mathcal{K} \bigcap M$ we have

$$(f, h) = (P^* g, h) = (g, Ph) = (g, h) = 0\,.$$

Hence, $f \in \mathcal{K}_* \bigcap (\mathcal{K} \bigcap M)^\perp$ and $\operatorname{clos} P^* M^\perp \subset \mathcal{K}_* \bigcap (\mathcal{K} \bigcap M)^\perp$.

Conversely, let $f \in \mathcal{K}_* \bigcap (\mathcal{K} \bigcap M)^\perp$. Then $P^* f = f$ and $f \in (\mathcal{K} \bigcap M)^\perp = \mathcal{K}^\perp \vee M^\perp$. Hence, $f = \lim_{n\to\infty}(k_n + m_n)$, where $k_n \perp \mathcal{K}$ and $m_n \perp M$. Since $\forall g \in \mathcal{H} \;\; (P^* k_n, g) = (k_n, Pg) = 0$ and therefore $P^* k_n = 0$, we have

$$\begin{aligned} f = P^* f &= P^*(\lim_{n\to\infty}(k_n + m_n)) \\ &= \lim_{n\to\infty}(P^* k_n + P^* m_n) \\ &= \lim_{n\to\infty} P^* m_n \in \operatorname{clos} P^* M^\perp\,. \end{aligned}$$

Hence, $\mathcal{K}_* \bigcap (\mathcal{K} \bigcap M)^\perp \subset \operatorname{clos} P^* M^\perp$. □

Theorem 3.6. $\mathcal{K}_{\text{out}} = \operatorname{clos} P_\theta \operatorname{Ran}(I - \pi_- \pi_-^\dagger)$.

Proof. Let $\theta = \theta_2\theta_1$ be the $*$-inner-outer factorization factorization of $\theta \in S_\Xi$ and (π_+, π_-) be the corresponding functional model. Let π be the operator corresponding (see (F&I)) to the invariant subspace $H_1 = \mathcal{K}_{*\,\text{in}}$. Taking in Lemma 3.4 $P = P_\theta$, $P_1 = P_{\theta_1}$, $P_2 = P_{\theta_2}$, we get $H_{2*} = \operatorname{Ran} P_2^* = \mathcal{K}_* \bigcap H_1^\perp$. On the other hand, $\mathcal{K}_{*\,\text{in}} = \mathcal{K} \bigcap M$, where $M = \operatorname{Ker}(I - \pi_+ \pi_+^\dagger)$. Then, by Lemma 3.5, we have $H_{2*} = \operatorname{Ran} P_2^* = \operatorname{clos} P^* M^\perp$. Thus we arrive at the description for $\mathcal{K}_{*\,\text{out}}$

$$\begin{aligned} H_{2*} = \operatorname{Ran} P_2^* &= \operatorname{clos} P^* \operatorname{Ran}(I - \pi_+ \pi_+^\dagger) \\ &= \operatorname{clos} P^* \operatorname{Ran}(I - \pi_{*-} \pi_{*-}^\dagger)\,, \end{aligned}$$

but for the dual model (π_{*+}, π_{*-}), where the operators $\pi_{*\pm}$ are uniquely determined by conditions $(f, \pi_{*\mp} v)_{\mathcal{H}} = < \pi_\pm^\dagger f, v >_C\,,\; f \in \mathcal{H},\; v \in L^2(\bar{C}, \mathfrak{N}_\mp)$. Taking into account the identity $P_\theta^* = P_{\theta_*}$, we can write the description for the terms of the original model $\mathcal{K}_{\text{out}} = \operatorname{clos} P_\theta \operatorname{Ran}(I - \pi_- \pi_-^\dagger)$. □

Remark. The invariant subspaces

$$\mathcal{K}_{*\,\mathrm{in}} = \{f \in \mathcal{K}_\theta : (I - \pi_+\pi_+^\dagger)f = 0\}$$

and

$$\mathcal{K}_{\mathrm{out}} = \operatorname{clos} P_\theta \operatorname{Ran}(I - \pi_-\pi_-^\dagger)$$

play important role for spectral analysis (see, e.g., [15]). The subspace $\mathcal{K}_{*\,\mathrm{in}}$ is called *singular* and $\mathcal{K}_{\mathrm{out}}$ is called *absolutely continuous*. This terminology descends from self-adjoint theory. To clarify it we define the spectral components [6]

$$M(T) = \{\, f \in H : \forall g \in H \quad ((T-z)^{-1}f, g)_+ = ((T-z)^{-1}f, g)_- \,\}$$

and

$$N(T) = \operatorname{clos}\left\{\, f \in H : \forall g \in H \quad ((T-z)^{-1}f, g) \in E^2\left(G_+ \bigcup G_-\right) \right\},$$

where $((T-z)^{-1}f, g)_\pm$ are the boundary limits of $((T-z)^{-1}f, g)$ from $G_\pm$. In the case when the operator T is self-adjoint (or unitary) we have $H_{sing}(T) = M(T)$ and $H_{ac}(T) = N(T)$. In the case when T is the main operator in functional model and $\Theta(z)^{-1}$ possesses boundary values a.e. on C, we have $\mathcal{K}_{*\,\mathrm{in}} = M(T)$ and $\mathcal{K}_{\mathrm{out}} = N(T)$.

References

[1] Garnett J. *Bounded analytic functions.* Academic Press, N.Y.-London, 1981.

[2] Szökefalvi-Nagy B. and Foiaş C. *Harmonic analysis of operators on Hilbert space.* North-Holland, Amsterdam-London, 1970.

[3] Koosis P. *Lectures on H_p spaces.* Cambridge Univ. Press, London, 1980.

[4] Nikolski N.K. *Operators, functions, and systems: an easy reading.* Vol. 1 Hardy, Hankel, and Toeplitz. Vol. 2 Model operators and systems, Math. Surveys and Monographs, 92, 93, AMS, Providence, RI, 2002.

[5] Lax P. and Phillips R.S. *Scattering theory.* Academic Press, NY, 1967.

[6] Tikhonov A.S. *Functional model and duality of spectral components for operators with continuous spectrum on a curve.* Algebra i Analiz, **14** (2002), no. 4, 158–195; English transl., St. Petersburg Math. J. **14** (2003), no.4, 169–184.

[7] Yakubovich D.V., *Linearly similar model of Sz.-Nagy–Foias type in a domain.* Algebra i Analiz, **15** (2003), no. 2, 180–227; English transl., St. Petersburg Math. J. **15** (2004), no. 2, 289–321.

[8] Tikhonov A.S. *Inner-outer factorization of J-contractive valued functions.* Operator Theory: Adv. and Appl., Vol. 118 (2000), 405–415.

[9] Beurling A., *On two problems concerning linear transformations in Hilbert space,* Acta Math. **81** (1949), 239–255.

[10] Livsic M.S., *On a class of linear operators on Hilbert space,* Mat. Sb. **19** (1946), 239–260.

[11] Tikhonov A.S. *On connection between factorizations of weighted Schur function and invariant subspaces.* Operator Theory: Adv. and Appl., Vol. 174 (2007), 205–246.

[12] Duren P.L., *Theory of H^p spaces*, Pure Appl. Math., vol. 38, Academic Press, New York–London, 1970.

[13] Abrahamse M.B. and Douglas R.G. *A class of subnormal operators related to multiply connected domains*, Adv. in Math., **19** (1976), 106–148.

[14] Tikhonov A.S. *Weighted Schur classes and a functional model.* (Russian) Fundam. Prikl. Mat. **12** (2006), no. 5, 221–236; translation in J. Math. Sci. (N.Y.) **150** (2008), no. 6, 2609–2619

[15] Pavlov B.S., *On conditions of separability of spectral components for a dissipative operator*, Izv. AN SSSR. Ser. mat. **39** (1975), no. 1, 123–148.

Alexey Tikhonov
Department of Mathematics
Taurida National University
4, Yaltinskaya str.
Simferopol 95007 Crimea, Ukraine
e-mail: tikhonov@club.cris.net

Operator Theory:
Advances and Applications, Vol. 214, 135–169
© 2011 Springer Basel AG

Eigenvalue Asymptotics for Magnetic Fields and Degenerate Potentials

Françoise Truc

Abstract. We present various asymptotic estimates of the counting function of eigenvalues for Schrödinger operators in the case where the Weyl formula does not apply. The situations treated seem to establish a similarity between magnetic bottles (magnetic fields growing at infinity) and degenerate potentials, and this impression is reinforced by an explicit study in classical mechanics, where the classical Hamiltonian induced by an axially symmetric magnetic bottle can be seen as a perturbation of the Hamiltonian derived from an operator with a degenerate potential.

Mathematics Subject Classification (2000). Primary 35P20.

Keywords. Spectrum, Min-max principle, Weyl asymptotics, Schrödinger operators, magnetic bottles.

1. Introduction

In this review are presented several results of spectral analysis, based for most of them on the min-max variational principle. These are mainly "non-Weyl-type" asymptotics for some "non generic" Schrödinger operators. In the appropriate setup, the Weyl formula describes the asymptotic relationship between the number of eigenvalues less than some fixed value λ and the volume, in phase space, of trajectories with energy less than λ for the corresponding classical problem. To be more precise, let us consider a continuous positive-valued potential V on $\mathbb{R}^m$, and let us make the following assumption for $V(x)$:

$$V(x) \to +\infty \quad \text{when} \quad |x| \to +\infty \tag{1.1}$$

(we call such a $V(x)$ a non degenerate potential). Then for any value of the parameter h in $]0,1]$, the operator $H_h = -h^2\Delta + V$ defined on $L^2(\mathbb{R}^m)$ is essentially self-adjoint and has a compact resolvent [49]. Moreover, denoting by $N(\lambda, H_h)$ the number of eigenvalues less than some fixed value λ, we get the following semi-classical asymptotic behaviour, when $h \to 0$:

$$N(\lambda, H_h) \sim h^{-m}(2\pi)^{-m} v_m \int_{\mathbb{R}^m} (\lambda - V(x))_+^{m/2} dx \,. \tag{1.2}$$

In this so-called semi-classical Weyl asymptotic formula, v_m denotes the volume of the unit ball in $\mathbb{R}^m$, and by W_+ we mean that we take the positive part of W.

If we take $h = 1$ in the previous formula we get the asymptotics for large energies of the operator $H_1 = -\Delta + V$:

$$N(\lambda, H_1) \sim_{\lambda \to +\infty} (2\pi)^{-m} v_m \int_{\mathbb{R}^m} (\lambda - V(x))_+^{m/2} dx \,. \tag{1.3}$$

The right-hand side of the formula (1.2) can be seen more generically as the volume, in phase space, of the set $\{(x,\xi), \mathcal{H}(x,\xi) \le \lambda\}$, where $\mathcal{H}(x,\xi) = \xi^2 + V(x)$ is the principal symbol of H_h and the Hamiltonian of the associated dynamics.

A naturel question is then the following: what can be said of a Schrödinger operator which has a discrete spectrum but does not verify the non-degeneracy condition (1.1)? In that case the volume of $\{(x,\xi), \xi^2 + V(x) \le \lambda\}$ may happen to be infinite, so that the formula (1.2) becomes irrelevant. This is the case for instance for the following potential (in $\mathbb{R}^2$)

$$V(x,y) = (1+x^2)y^2$$

(see Figure 1).

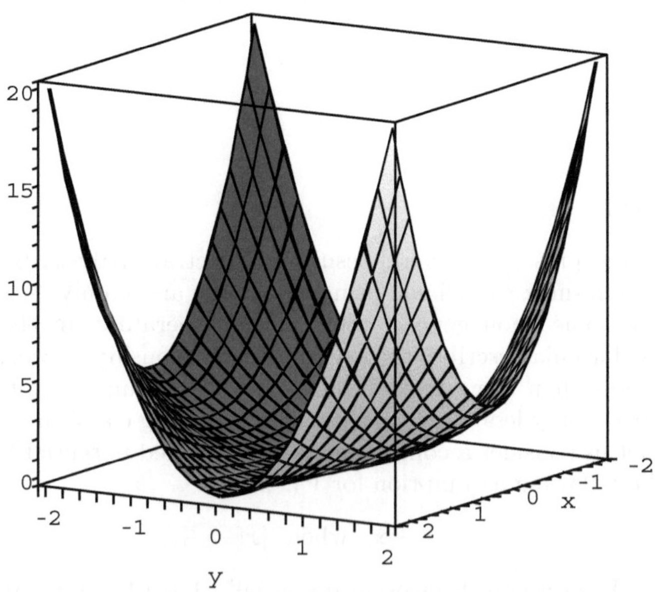

FIGURE 1. The potential $V(x,y) = (1+x^2)\, y^2$.

The problems presented below discuss precisely this question for various situations, and the estimates obtained will be called non-Weyl-type asymptotics.

First we recall the results obtained in ([39], [41]), for a class of degenerate potentials in the sense we previously defined. These potentials can be seen as a generalization of the preceding example; they are of the form

$$V(x) = f(y)g(z), \quad x = (y, z) \in \mathbb{R}^n \times \mathbb{R}^p,$$
$$f \in C(\mathbb{R}^n; \mathbb{R}_+^*) \quad g \in C(\mathbb{R}^p; \mathbb{R}_+),$$
$$g \text{ homogeneous of degree } a.$$

Then we consider Schrödinger operators with magnetic field $H_h(A) = ((h\nabla - iA))^2$. One can call them degenerate in the sense that the principal symbol of $H_h(A)$, which is $\mathcal{H}(x,\xi) = (\xi - A(x))^2$, annihilates on a non compact manifold of $T^*(\mathbb{R}^m)$. If the magnetic field $B = dA$ is such that the counting function $N(\lambda, H_h(A))$ can be defined, then we can look for some alternative to Weyl formula. In particular, when the magnetic field $B = dA$ satisfies some so-called magnetic bottles conditions:

$$\|B(x)\| \to +\infty \quad \text{if} \quad |x| \to +\infty\,, \tag{1.4}$$

$H_h(A)$ is essentially self-adjoint and has a compact resolvent on $L^2(\mathbb{R}^m)$ [3]. The spectral asymptotics for large energies were computed by Y. Colin de Verdière [6]. Here are discussed the semi-classical version of this result [59], and the case of magnetic bottles in the hyperbolic context ([42] for the Poincaré half-plane, [43] for geometrically finite hyperbolic surfaces). These non-Weyl-type asymptotics can be seen as the expression of an integrated density of states on the whole space. For a constant magnetic field $B = \sum_{j=1}^r b_j dx_j \wedge dy_j,\ \ b_1 \geq b_2 \geq \cdots \geq b_r > 0$, the density of states is given by, (for some universal constant C_r):

$$\nu_B(\lambda) = C_r b_1 b_2 \cdots b_r \sum_{n_j \geq 0} \left(\lambda - \sum_{j=1}^r (2n_j + 1) b_j\right)_+^{d/2-r}.$$

In the hyperbolic context $r(x) = 1$ for any x and the intensity $b(x)$ is defined in a slightly different way according to the hyperbolic geometry.

We discuss also another "degenerate" problem in the framework of the superconductivity theory. In order to minimize the associated Ginzburg-Landau functional associated to a given open set Ω in $\mathbb{R}^3$, we study the spectral properties of the magnetic Laplacian $H = ((\nabla - iA))^2$, with the so-called magnetic Neumann condition at the boundary:

$$\nu(x) \cdot (\nabla - iA)u(x) = 0 \quad \forall x \in \delta\Omega\,.$$

This comes from the fact that $(0, \sigma A)$ is a trivial critical point for the functional (σ is a parameter related to the magnetic intensity); the magnetic Laplacian previously defined is precisely the Hessian computed at this point.

The magnetic field $B = dA$ is assumed to be constant, and the spectrum contains an absolutely continuous part, which is the whole interval $[b, +\infty[$. ($b =$

$\|B\|$ is the magnetic intensity). However, if we consider the case of the half-space, e.g., for (t, x, y) in $\Omega = \mathbb{R}_+ \times \mathbb{R}^2$, the Neumann realization of the magnetic Laplacian

$$H = (D_t - A_1)^2 + (D_x - A_2)^2 + (D_y - A_3)^2,$$

with $D_s = -i(\frac{\partial}{\partial s})$, and if we assume that the magnetic field is not orthogonal to the boundary $\delta\Omega = \{(0, x, y), (x, y) \in \mathbb{R}^2\}$, we get that the lower bound of the spectrum is strictly less than b, [36], [24]. It can be proved that the part of the spectrum less than b consists of a finite number of eigenvalues, each one having an infinite multiplicity [40].

Furthermore, this number tends to infinity as the angle between the magnetic field and the boundary $\delta\Omega$ tends to zero, and this leads to a non-Weyl-type description of the number of eigenvalues less than a fixed real number less than b [40].

The last section is devoted to a problem of magnetic bottle, but in the classical context [58]. This problem is indeed at the origin of the results proved in the semi-classical and quantum mechanics context. The magnetic field considered here is axially symmetric in $\mathbb{R}^3$ and verifies the condition (1.4). Are there bounded trajectories, as suggested by numerical simulations? The operator associated to the Hamiltonian by the Weyl quantification has a discrete spectrum: we are in the case of the magnetic bottles defined previously. However in the classical setup we have to use the results of the KAM theory. The conditions needed to apply Moser's twist theorem have to be checked and then we can conclude that there exists an open set of initial conditions such that the trajectory is bounded. The interesting fact is that the Hamiltonian can be described as a perturbation of an effective Hamiltonian, which is precisely the principal symbol of the Schrödinger operator with the degenerate potential

$$V(x, y) = B^2(x, 0)\, y^2 \,.$$

2. Degenerate potentials

2.1. The Tauberian approach

There are a lot of works on the subject, and we refer to [60] for a review. However for the reader's convenience we recall briefly the main results in this approach. Roughly speaking, the Tauberian technique consists on studying the asymptotic behaviour of the Green's function of the operator H_1 and applying a Tauberian theorem. [12] is the first result where (1.3) is proved for a class of non degenerate potentials, then refinements can be found in [56], [33], [31] and [51], where the formula (1.3) is proved under minimal conditions on V.

In [53] Solomyak makes the following remark:

Lemma 2.1. *Let V be a positive a-homogeneous potential:*

$$V(x) \geq 0; \quad V(tx) = t^a V(x) \text{ for any } t \geq 0 \ (a > 0).$$

If moreover $V(x)$ is strictly positive ($V(x) \neq 0$ if $x \neq 0$) the spectrum of H_1 is discrete and the formula (1.3) *takes the form:*

$$N(\lambda, H_1) \sim \gamma_{m,a} \lambda^{\frac{2m+am}{2a}} \int_{S^{m-1}} (V(x))^{-m/a} dx \tag{2.1}$$

($\gamma_{m,a}$ *is a constant depending only on the parameters* m *and* a.)

From that lemma comes out naturally the idea of investigating the spectrum without the condition of strict positivity (and thus in a case of degeneracy of the potential); the two main results are [53]:

Theorem 2.2. *The formula* (2.1) *still holds for a positive a-homogeneous potential such that $J(V) = \int_{S^{m-1}} (V(x))^{-m/a} dx$ is finite.*

The second result deals with a case where $J(V)$ is infinite:

Theorem 2.3. *Let $V(x) = F(y,z)$, $y \in \mathbb{R}^n$, $z \in \mathbb{R}^p$, $n+p=m$, $m \geq 2$, such that $F(sy,tz) = s^b t^{a-b} F(y,z)$ (with $0 < a < b$) and $F(y,z) > 0$ for $|z||y| \neq 0$. Denote by $\lambda_j(y)$ the eigenvalues of the operator $-\Delta_z + F(y,z)$ in $L^2(\mathbb{R}^p)$ and let $s = \frac{2b}{2+a-b}$, then:*

$$\begin{aligned} \text{If } \frac{n}{b} > \frac{m}{a} \quad N(\lambda, H_1) &\sim \gamma_{n,s} \lambda^{\frac{2m+am}{2b}} \int_{S^{m-1}} \Sigma(\lambda_j(y))^{-n/s} dx, \\ \text{if } \frac{n}{b} = \frac{m}{a} \quad N(\lambda, H_1) &\sim \frac{a(a+2)}{2b(a-b)} \gamma_{m,a} \lambda^{\frac{2m+am}{2b}} \ln \lambda \int_{S^{n-1} S^{p-1}} F(y,z)^{-m/a} dx. \end{aligned}$$

The proof is based on variational techniques and spectral estimates of [51].

In [50] D. Robert extends the theory of pseudodifferential operators to pseudodifferential operators with operator symbols. It is thus possible to study cases where the operator has a compact resolvent but the condition $\lim_\infty V(x) = +\infty$ is not fulfilled. As an example it gives the asymptotics of $N(\lambda, H_1)$ for the two-dimensional potential $V(y,z) = y^{2k}(1+z^2)^l$, where k and l are strictly positive. The asymptotics are the following:

Theorem 2.4.

$$\begin{aligned} \text{If } k > l \quad N(\lambda, H_1) &\sim \gamma_1 \lambda^{\frac{l+k+1}{2l}}, \\ \text{if } k = l \quad N(\lambda, H_1) &\sim \gamma_2 \lambda^{\frac{2k+1}{2k}} \ln \lambda, \\ \text{if } k < l \quad N(\lambda, H_1) &\sim \gamma_3 \lambda^{\frac{2k+1}{2k}}. \end{aligned}$$

The constants γ_i depend only on k and l, but the first one γ_1 takes into account the trace of the operator $(-\Delta_z + z^{2k})^{-(k+1)/2l}$ in $L^2(\mathbb{R})$.

In the two-dimensional case let us mention the results of B. Simon [52]. He first recalls Weyl's famous result: let H be the Dirichlet Laplacian in a bounded domain Ω in $\mathbb{R}^2$, then the following asymptotics hold:

$$N(\lambda, H) \sim \frac{1}{2} \lambda |\Omega|$$

and then he considers domains Ω for which the volume (denoted by $|\Omega|$) is infinite but the spectrum of the Laplacian is still discrete. These domains are of the type

$$\Omega_\mu = \{(y,z); |y||z|^\mu \leq 1\}.$$

Actually the problem can be derived from the study of the asymptotics of Schrödinger operators with the homogeneous potential: $V(y,z) = |y|^\alpha |z|^\beta$.

In order to compute eigenvalue asymptotics, he uses the Feynman-Kac formula and the Karamata-Tauberian theorem, but the main tool is what he calls "sliced bread inequalities", which can be seen as a kind of Born-Oppenheimer approximation. More precisely let $H = -\Delta + V(y,z)$ be defined on $\mathbb{R}^{n+p}$, and denote by $\lambda_j(y)$ the eigenvalues of the operator $-\Delta_z + V(y,z)$ in $L^2(\mathbb{R}^p)$. (If the z's are electron coordinates and the y's are nuclear coordinates, the $\lambda_j(y)$ are the Born-Oppenheimer curves). He proves the following lemma:

$$\mathrm{Tr} e^{-tH} \leq \Sigma_j e^{-t(-\Delta_y + \lambda_j(y))}$$

(when the second term exists).

Thus he gets the two following coupled results:

Theorem 2.5. *If $H = -\Delta + |y|^\alpha |z|^\beta$ and $\alpha < \beta$, then*

$$N(\lambda, H) \sim c_\nu \lambda^{\frac{2\nu+1}{2}} \qquad \left(\nu = \frac{\beta+2}{2\alpha}\right)$$

Corollary 2.6. *If $H = -\Delta_{\Omega_\mu}$ $(\mu > 1)$, then $N(\lambda, H) \sim c_\mu \lambda^{\frac{1}{2\mu+1}}$.*

Theorem 2.7. *If $H = -\Delta + |y|^\alpha |z|^\alpha$, then $N(\lambda, H) \sim \frac{1}{\pi} \lambda^{1+\frac{1}{\alpha}} \ln \lambda$.*

Corollary 2.8. *If $H = -\Delta_{\Omega_\mu}$ $(\mu = 1)$, then $N(\lambda, H) \sim \frac{1}{\pi} \lambda \ln \lambda$.*

The constant c_μ depends only on μ, and the constant c_μ takes in account the trace of the operator $(-\Delta_z + |z|^\beta)^{-\nu}$ in $L^2(\mathbb{R})$.

2.2. The min-max approach

The result presented in this section [39] is based on the method of Courant and Hilbert, the min-max variational principle.

Thanks to this method, which requires only to study the associated quadratic form, (using appropriate partitions and simplified models) no assumptions on the evolution semi-group are needed, and we get non-Weyl-type asymptotics for a large class of degenerate potentials, namely potentials of the following form:

$$\begin{aligned} &V(x) = f(y)g(z), \quad x = (y,z) \in \mathbb{R}^n \times \mathbb{R}^p, \quad n+p = m, \quad m \geq 2 \\ &f \in C(\mathbb{R}^n; \mathbb{R}_+^*), \quad g \in C(\mathbb{R}^p; \mathbb{R}_+), \\ &\exists\ a > 0 \ \text{ t.q. } \ g(tz) = t^a g(z) \ \ \forall t > 0, \quad g(z) > 0 \ \forall z \neq 0. \end{aligned} \tag{2.2}$$

This class contains the potentials studied in [50], [52] and [53].

According to the assumption (2.2) the spectrum of the operator $-\Delta_z + g(z)$ on $L^2(\mathbb{R}^p)$ is discrete and positive. Let us denote by μ_j its eigenvalues.

We have moreover:

Remark 2.9. If $f(y) \to +\infty$ when $|y| \to +\infty$, then $H_h = -h^2\Delta + V$ has a compact resolvent.

Of course if f was assumed to be homogeneous, the asymptotics would be given by Theorem 2.3. But here the only additional assumption made on f is a locally uniform regularity:

$$\begin{aligned} &\exists\, b, c > 0 \;\; t.q. \;\; c^{-1} \leq f(y) \qquad \text{and} \\ &|f(y) - f(y')| \leq \; cf(y)|y - y'|^b, \;\; \forall (y, y') \; t.q. \; |y - y'| \leq 1. \end{aligned} \tag{2.3}$$

Theorem 2.10. *Let us assume the previous conditions on f and g. Then there exists $\sigma, \tau \in]0,1[$ such that, for any $\lambda > 0$, one can find $h_0 \in]0,1[$, $C_1, C_2 > 0$ in order to have*

$$(1 - h^\sigma C_1) n_{h,f}(\lambda - h^\tau C_2) \leq N(\lambda; H_h) \leq (1 + h^\sigma C_1) n_{h,f}(\lambda + h^\tau C_2) \;\; \forall h \in]0, h_0[$$

with $n_{h,f}(\lambda) = h^{-n}(2\pi)^{-n} v_n \int_{\mathbb{R}^n} \Sigma_{j\in\mathbf{N}} [\lambda - h^{2a/(2+a)} f^{2/(2+a)}(y)\mu_j]_+^{n/2} dy.$

Provided some additional conditions on f, the previous result can be refined as follows:

Theorem 2.11. *If moreover one can find a constant C_3 such that, for any $\mu > 1$:*

$$\int_{\{y, f(y) < 2\mu\}} f^{-p/a}(y) dy \leq C_3 \int_{\{y, f(y) < \mu\}} f^{-p/a}(y) dy \, ,$$

then one can take $C_2 = 0$ in Theorem 2.10:

$$(1 - h^\sigma C_1) n_{h,f}(\lambda) \leq N(\lambda; H_h) \leq (1 + h^\sigma C_1) n_{h,f}(\lambda) \quad \forall h \in]0, h_0[\, .$$

Remark 2.12. If $f^{-p/a} \in L^1(\mathbb{R}^n)$ and $g \in C^1(\mathbb{R}^p \backslash \{0\})$, then the formula (1.3) holds.

The proof of Theorem 2.10 is based on a suitable subdivision of $\mathbb{R}^n$ into cubes $\{Q_r(r\gamma), \gamma \in \mathbb{Z}^n\}$. According to the min-max variational principle we are then reduced to study Dirichlet and Neumann problems in cylinders of $\mathbb{R}^m$:

$$\begin{aligned} N(\lambda, H^D_{h,\gamma}) \leq N(\lambda, H_h) &\leq N(\lambda, H^N_{h,\gamma}) \\ H^{D,N}_{h,\gamma} = -h^2\Delta_y - h^2\Delta_z &+ f(y)g(z) \quad \text{on} \;\; Q_r(r\gamma) \times \mathbb{R}^p \, , \end{aligned}$$

with Dirichlet (or Neumann) condition at the boundary.

In each cube $Q_r(r\gamma)$ $f(y)$ is bounded from above by $f(y^*_\gamma)$ where y^*_γ is a minimum for f.

Using the homogeneity of g one gets that the eigenvalues of the operator $-h^2\Delta_z + f(y^*_\gamma)\, g(z)$ are of the form $\{(h^a f(y^*_\gamma))^\alpha \; \mu_j \; \}$ ($\alpha = 2/(2+a)$).

One gets then a lower bound for $N(\lambda, H_h)$ by taking the sum, for all cubes, of the sum for all j's of $N\big(\lambda - (h^a f(y^*_\gamma))^\alpha \; \mu_j, \; -h^2\Delta^D_{Q_r(r\gamma)}\big)$.

One gets the upper bound following the same procedure.

Concerning the proof of Theorem 2.11, the main tool is an asymptotic formula of the moment of eigenvalues of $-h^2\Delta_z + g(z)$, which is again obtained using the min-max principle.

As a conclusion, let us notice that if there is some information on the growth of f at infinity, then the asymptotics can be computed in terms of power of h:

Remark 2.13. If there exist $k > 0$ and $C > 0$ such that

$$\frac{1}{C}|y|^k \leq f(y) \leq C|y|^k \text{ for } |y| > 1,$$

then

$$\begin{array}{ll} \text{if } \; k > a & N(\lambda, H_h) \approx h^{-m} \\ \text{if } \; k = a & N(\lambda, H_h) \approx h^{-m} \ln \dfrac{1}{h} \\ \text{if } \; k < a & N(\lambda, H_h) \approx h^{-n-\frac{pa}{k}} \end{array}$$

Remark 2.14. The formula in Theorem 2.10 gives us a hint of what can be said about the behaviour of the eigenvalues themselves. Actually this question is answered precisely by using Born-Oppenheimer-type methods in [41], where we compute first-order approximations for low energies and middle energies, and apply the results to a potential vanishing on a hypersurface.

3. Magnetic bottles

3.1. General setting

We are now interested in magnetic Laplacians, in various situations when it is possible to look for non-Weyl-type estimates. This leads us to give a definition of magnetic bottles in a general Riemannian context.

Let us denote by (M, g) a connected Riemannian manifold of dimension d and by $A = \sum_{j=1}^{d} a_j dx_j$ a real one-form on M. For any $h \in]0, 1[$ we can define the semi-classical magnetic Laplacian

$$\begin{aligned} H_h(A) &= (ih\, d + A)^\star (ih\, d + A)\,, \\ (ih\, d + A)u &= ih\, du + Au\,,\ \forall\, u\, \in\, C_0^\infty(M)\,. \end{aligned} \tag{3.1}$$

The magnetic field is the exact two-form $B = dA$.

The two-form B is associated to a linear operator L_B on the tangent space defined by

$$B(X, Y) \;=\; g(L_B.X, Y)\,; \quad \forall\, X\,,\; Y\, \in TM \times TM\,. \tag{3.2}$$

The magnetic intensity $\mathbf{b}$ is given by

$$\mathbf{b} \;=\; \frac{1}{2}\mathrm{tr}\left((B^\star B)^{1/2}\right)\,. \tag{3.3}$$

It is possible to define $H_h(A)$ more geometrically, using the Hermitian connection ∇ on a complex-line bundle L over M with curvature equal to iB. This connection exists provided that the cohomology class of $B/2\pi$ is an integer.

It is defined by $\nabla_X f = df(X) - iA(X)f$, where A is a real one-form verifying $B = dA$. One introduces on $C_0^\infty(M;L)$ the quadratic form $q(f) = \int_M \|\nabla f\|^2 dx$, and by Friedrich's process one gets an operator, which is $H_h(A)$.

Remark 3.1 (Gauge invariance). If $A' = A + d\phi$ is another magnetic potential associated to B, the operators $H_h(A)$ and $H_h(A')$ are unitarily equivalent.

This property implies that $H_h(A)$ and $H_h(A')$ have the same spectrum. Therefore we give the following definition, which does not depend from the choice of the magnetic potential A:

Definition 3.2. (M, h, B) is called a magnetic bottle if

1) $H_h(A)$ is essentially self-adjoint with domain $C_0^\infty(M;L)$,
2) $H_h(A)$ has a compact resolvent.

In [3], which is the first paper on the subject, and also in [14], [30] one can find necessary conditions or sufficient conditions for $H_h(A)$ to have a compact resolvent.

3.2. The Euclidean case

3.2.1. The results. Let us take for (M, g) the Euclidean space $\mathbb{R}^d$. The operator defined in (3.1) is

$$H_h(A) = \sum_{j=1}^{d} \left(\frac{h}{i} \frac{\partial}{\partial x_j} - a_j \right)^2 .$$

Furthermore there exists, for any $x \in \mathbb{R}^d$, an orthonormal basis $(e_j(x))$ of $\mathbb{R}^d$ such that $B(x)$ has the following expression

$$B(x) = \sum_{j=1}^{r(x)} b_j(x) dx_j \wedge dy_j, \quad b_1(x) \geq b_2(x) \geq \cdots \geq b_r(x) > 0 . \tag{3.4}$$

The magnetic intensity is equal to the norm of the vector $B(x) = (b_j(x))_j$. The $b_j(x)$ are the moduli of the non zero eigenvalues of the endomorphism L_B associated to $B(x)$ and $2r(x)$ is the rank of L_B. For odd dimension in particular 0 is always an eigenvalue. We assume moreover the following properties for B:

(B_1) $\lim_{\|x\| \to \infty} \|B(x)\| = \infty$,

(B_2) there exists $C > 0$ such that, for every x and x' verifying:

$$\|x - x'\| \leq 1, \quad \|B(x)\| \leq C \|B(x')\|,$$

(B_3) $M(x) = o\left(\|B(x)\|^{\frac{3}{2}} \right)$ when $\|x\| \to \infty$ where

$$M(x) = \max_{|\beta|=2} \left(\sup_{\|x-x'\| \leq 1} \|D^\beta A(x')\| \right) .$$

The high energy behaviour of $N(\lambda, H_1(A))$, $(h = 1, \lambda \to +\infty)$, is given by Y. Colin de Verdière in [6]:

Theorem 3.3. *Under the conditions* (B_1–B_3), $(\mathbb{R}^d, 1, B)$ *is a magnetic bottle and*

$$N_B^{as}[\lambda(1-o(1))] \leq N(\lambda, H_1(A)) \leq N_B^{as}[\lambda(1+o(1))] \quad (\lambda \to +\infty)\,.$$

The expression for N_B^{as} *is the following:*

$$N_B^{as}(\lambda) = \sum_{r=1}^{[d/2]} C_{k,r} \sum_{(n_1,\dots,n_r)\in Z_r^+} \int_{A_r} (\lambda - \sum_{i=1}^{r}(2n_i+1)b_i(x))_+^{k/2} \prod_{i=1}^{r} b_i(x)dx\,.$$

We used the following notations:

- $A_r = \{x \in R^d; r(x) = r\}$
- $C_{k,r} = \frac{\gamma_k}{(2\pi)^{k+r}}$, γ_k = volume of the unit ball of R^k.

In [59] we give an equivalent of $N(E, H_h(A))$ for a fixed energy E when h tends to zero. This is the semi-classical version of the previous asymptotics.

We first notice that $H_h(A) = h^2 H_1(A/h)$. $H_1(A/h)$ is the (non semi-classical) Schrödinger operator associated to the magnetic field $\frac{B}{h}$:

$$H_1(A/h) = \sum_{i=1}^{d} \left(\frac{1}{i}\frac{\partial}{\partial x_j} - \frac{a_j}{h}\right)^2 . \tag{3.5}$$

Consequently, we get $N(E, H_h(A)) = N\left(\frac{E}{h^2},\ H_1(A/h)\right)$ for any fixed energy E.

Using an adaptation of the method explained in [6], we get the following asymptotics [59]:

Theorem 3.4. *Under the conditions* (B_1–B_3), $(\mathbb{R}^d, h, B)$ *is a magnetic bottle and we have, for any energy* E*:*

$$\frac{1}{h^d} N_{hB}^{as}[E(1-o(1))] \leq N(E, H_h(A)) \leq \frac{1}{h^d} N_{hB}^{as}[E(1+o(1))] \quad (h \to 0)\,.$$

Remark 3.5. The expression for N_B^{as} becomes more explicit when $d = 2$. We have then

$$b_1(x) = \|B(x)\| = \mathbf{b}(x),$$

and

$$\frac{1}{h^d} N_{hB}^{as}(E) = \frac{1}{2\pi h^2} \int_{R^2} \mathbf{b}(x) \sum_{n\in\mathbb{N}} \left[E - (2n+1)h\mathbf{b}(x)\right]_0^+ dx\,. \tag{3.6}$$

$[\rho]_+^0$ is the Heaviside function:

$$[\rho]_+^0 = \begin{cases} 1\,, & \text{if} \quad \rho > 0 \\ 0\,, & \text{if} \quad \rho \leq 0\,. \end{cases}$$

Remark 3.6. H. Matsumoto recovers the conclusions of this theorem by studying the semi-group $\exp(-tH_h)$ [38]. The following equivalent is obtained:

$$\mathrm{Tr}\big(\exp(-tH_h)\big) = \frac{1}{h^d} Z_{hB}(t)$$

where $Z_{hB}(t) = (4\pi t)^{-d/2} \int_{R^d} \prod_{i=1}^{r(x)} \frac{htb_i(x)}{\sinh htb_i(x)} dx$ is the Laplace transform of the function $N_{hB}^{as}(\lambda)$ introduced previously. In the three-dimensional case Tamura [55] obtains a result of the same kind, involving only the norm of the magnetic field. However they both require stronger conditions for B, in order to make sure that $\exp(-tH_h)$ is a trace semi-group. This comes from the philosophy of the min-max method, which does not deal with the evolution semi-group but requires only to study the quadratic form, using partitions and asymptotic formulas for simplified operators (namely here for constant fields in cubes), so that we can get a formula with the minimal assumptions.

Remark 3.7. The operators verifying the assumptions of the theorem are special examples of the hypoelliptic operators $\sum_{i=1}^{d} X_k^* X_k$ introduced by L. Hörmander [25], in the case of real vector fields X_k.

Remark 3.8. Let us set

$$\nu_{B(x)}(\lambda) = C_{k,r} \sum_{(n_1,\dots,n_r)\in Z_r^+} \left(\lambda - \sum_{i=1}^{r}(2n_i+1)b_i(x)\right)_+^{k/2} \prod_{i=1}^{r} b_i(x) \, .$$

(In this definition, the numbers k and r depend on x.) The function $N_B^{as}(\lambda)$ has then the following expression:

$$N_B^{as}(\lambda) = \int_{R^d} \nu_{B(x)}(\lambda) dx \, .$$

In the case of a constant magnetic field, the function $\nu_B(\lambda)$ can be seen as a density of states for the Schrödinger operator in $\mathbb{R}^d$.

The proof consists of two main parts which we develop in next sections: the asymptotic spectral estimate for the Dirichlet problem in the cube $[0,R]^d$ in the case of a constant field, and the appropriate subdivision in cubes which makes possible the reduction to the simplified problem.

3.2.2. The Dirichlet problem in a cube for a constant magnetic field. When the field B is constant, the function $\nu_B(\lambda)$ is used to estimate $N_{B,R}(\lambda)$, the counting function of the spectrum concerning Dirichlet problem for the Schrödinger operator with the magnetic field B in the cube $[0,R]^d$. We recall the precise estimate, given in [6]:

Theorem 3.9. *There exists a constant c depending only on d such that, for any A with* $0 < A < R/2$, *the following inequalities hold:*

- $N_{B,R}(\lambda) \leq R^d \; \nu_B(\lambda)$.
- $N_{B,R}(\lambda) \geq (R-A)^d \; \nu_B(\lambda - C/A^2)$.

The proof of this result uses the spectrum for constant fields on a torus, and a method due to Polya, which consists in subdividing $\mathbb{R}^d$ into cubes and taking an approximation by a "large" torus ([6], [8] and [9]).

To be more precise we have:

Lemma 3.10 (Constant field on a torus). *Let* $B = \sum_{j=1}^{r} b_j dx_j \wedge dy_j$, $b_1 \geq \cdots \geq b_r > 0$ *on the torus* $M = \mathbb{R}^d/\Gamma$, *where* $d = r + k$ *and* Γ_0 *is a lattice on* $\mathbb{R}^k$. *It is assumed that* $\Gamma = \bigoplus_{j=1}^{r} \rho_j \mathbb{Z}^2 \, \oplus \Gamma_0$ *and that* $b_j \rho_j^2 \in 2\pi\mathbb{Z}$. *Then*

1) *the cohomology class of* $B/2\pi$ *is an integer,*
2) *the spectrum of* $H_h(B)$ *is constituted of the eigenvalues* $\lambda = \sum_{i=1}^{r}(2n_i+1)b_i + \mu$, *where* $n_i \in \mathbb{N}^*$, *and* μ *is an eigenvalue of the Laplacian on* $\mathbb{R}^k/\Gamma_0$,
3) *the multiplicity of* λ *is equal to the sum of the multiplicity of* μ *and of* $\prod_{i=1}^{r} \frac{b_i \rho_i^2}{2\pi}$.

3.2.3. A subdivision of R^d into appropriate cubes.

Lemma 3.11. *Under the assumptions* (B_1–B_3), *and for a fixed* $\varepsilon > 0$, *there exists for any* h *a subdivision of* R^d *in cubes* $(\Omega_i)_{i \geq 0}$ *of sides* r_i, *and numbers* $(a_i)_{i \geq 1}$ $(0 < a_i \leq r_i/2)$ *such that, if we set* $M_i = \max_{\|\beta\|=2} \sup_{x \in \Omega_i} \|D^\beta a(x)\|$, *the following inequalities hold, for any* x *in* Ω_i *and for any integer* $i \geq 1$:

i) $r_i^2 M_i \leq \varepsilon h \|B(x)\|^{1/2}$
ii) $M_i \leq \varepsilon^3 \|B(x)\|^{3/2}$
iii) $1/a_i^2 \leq M_{x,\varepsilon} = \max\left(\frac{4\varepsilon\|B(x)\|}{h}, 1/\varepsilon\right)$.

3.3. The hyperbolic half-plane

3.3.1. The setup. Now, we consider the case where $M = \mathbb{H}$ is the hyperbolic plane: $\mathbb{H} = \mathbb{R}\times]0,+\infty[$, endowed with the hyperbolic metric $g = \dfrac{dx^2 + dy^2}{y^2}$. We have $\rho_B = \widetilde{\mathbf{b}}\, dv$, where $dv = y^{-2} dx dy$ is the Riemannian measure on M. Thus we have $\widetilde{\mathbf{b}} = y^2 \left(\partial_x A_2 - \partial_y A_1\right)$ and

$$H_1(A) = y^2(D_x - A_1)^2 + y^2(D_y - A_2)^2 , \tag{3.7}$$

We define $\mathbf{b} = |\widetilde{\mathbf{b}}|$.

The hyperbolic framework has been used mainly for studying the Maass Laplacian, which corresponds to the constant magnetic field case. This case has been studied by many authors [21], [15], [11] [13]. In [27] Y. Inahama and S. Shirai consider asymptotically constant magnetic fields and in [29] they deal with Pauli operators. In [26] N. Ikeda studies the relationship between Maass Laplacian and Schrödinger operators with Morse potentials.

From an other point of view, the asymptotic distribution of large eigenvalues in the hyperbolic context has already been studied for Schrödinger operators (without magnetic field) [28]. The method is based on Feynman-Kac representation of the heat kernel and the Tauberian theorem. As already mentioned our own method involves only min-max techniques so it does not require to study properties of the evolution semigroup. We get the asymptotic distribution of large eigenvalues for a certain type of magnetic bottles following the method used in the Euclidean case, but replacing cubes by rectangles adapted to the hyperbolic geometry.

The result is very similar to (3.6), according to the hyperbolic definition of the intensity $\widetilde{\mathbf{b}}$ of the magnetic field. Moreover the techniques are local, so they have been successfully applied to geometrically finite hyperbolic surfaces of infinite area [43].

We give first the basic results in the case of a constant magnetic field.

3.3.2. The Maass Laplacian. The first paper on Maass Laplacian is due to J. Elstrodt [15].

We consider the case where $\widetilde{\mathbf{b}} = y^2(\partial_x A_2(x,y) - \partial_y A_1(x,y))$ is constant. We choose a gauge such that $A_2 = 0$, so $A_1(x,y) = \widetilde{\mathbf{b}}y^{-1}$. We can assume that $A_1(x,y) = \mathbf{b}y^{-1}$, by eventually performing the change $x \to -x$, which is a unitary operator on $L^2(\mathbb{H})$. The operator we are interested in is

$$H_1 A^{\mathbf{b}} = y^2(D_x - \mathbf{b}y^{-1})^2 + y^2 D_y^2\,, \quad \text{with} \quad \mathbf{b} \geq 0 \quad \text{constant.} \tag{3.8}$$

Let U be the unitary operator

$$U : L^2(\mathbb{H}) \to L^2(\mathbb{R}\times\mathbb{R}_+)\,, \quad Uf = y^{-1}f\,; \tag{3.9}$$

$\mathbb{R}\times\mathbb{R}_+$ is endowed with the standard Lebesgue measure $dxdy$. Then

$$P_{\mathbf{b}} = U(-\Delta_{A^{\mathbf{b}}})U^\star = (D_x - \mathbf{b}y^{-1})y^2(D_x - \mathbf{b}y^{-1}) + D_y y^2 D_y\,. \tag{3.10}$$

Using partial Fourier transform we get that $\mathrm{sp}(P_{\mathbf{b}}) = \bigcup_{\xi\in\mathbb{R}} \mathrm{sp}(P_{\mathbf{b}}(\xi))$, where $P_{\mathbf{b}}(\xi)$ is the self-adjoint operator on $L^2(\mathbb{R}_+)$ defined by

$$P_{\mathbf{b}}(\xi)f = (y\xi - \mathbf{b})^2 f(y) + D_y(y^2 D_y f)(y)\,; \quad \forall\, f \in C_0^\infty(\mathbb{R}_+)\,. \tag{3.11}$$

Moreover we have

$$\mathrm{sp}(P_{\mathbf{b}}(\xi)) = \mathrm{sp}(P_{\mathbf{b}}(1))\,, \quad \text{if} \quad \xi > 0\,.$$
$$\mathrm{sp}(P_{\mathbf{b}}(\xi)) = \mathrm{sp}(P_{\mathbf{b}}(-1))\,, \quad \text{if} \quad \xi < 0\,.$$

This leads to the well-known following theorem:

Theorem 3.12. *The spectrum of $P_{\mathbf{b}}(\pm 1)$ is formed by its absolutely continuous part and its discrete part, and*

$$\mathrm{sp}(P_{\mathbf{b}}(-1)) = \mathrm{sp}_{\mathrm{ac}}(P_{\mathbf{b}}(-1)) = \mathrm{sp}_{\mathrm{ac}}(P_{\mathbf{b}}(1)) = \Big[\mathbf{b}^2 + \frac{1}{4}, +\infty\Big[$$
$$\mathrm{sp}(P_{\mathbf{b}}(1)) = \mathrm{sp}_{\mathrm{ac}}(P_{\mathbf{b}}(1)) \quad , \quad \text{if } \mathbf{b} \leq \frac{1}{2}$$
$$\mathrm{sp}_d(P_{\mathbf{b}}(1)) = \Big\{(2j+1)\mathbf{b} - j(j+1)\,;\ j \in \mathbb{N}\,,\ j < \mathbf{b} - \frac{1}{2}\Big\} \quad \text{if } \mathbf{b} > \frac{1}{2}\,.$$

Corollary 3.13. *The spectrum of $-\Delta_{A^{\mathbf{b}}}$ is essential:* $\mathrm{sp}(-\Delta_{A^{\mathbf{b}}}) = \mathrm{sp}_{\mathrm{es}}(-\Delta_{A^{\mathbf{b}}})$. *Its absolutely continuous part is given by* $\mathrm{sp}_{\mathrm{ac}}(-\Delta_{A^{\mathbf{b}}}) = \big[\mathbf{b}^2 + \frac{1}{4}, +\infty\big[$. *The remaining part of its spectrum is empty if $0 \leq \mathbf{b} \leq 1/2$, otherwise it is formed by a finite number of eigenvalues of infinite multiplicity given by*

$$\mathrm{sp}_p(-\Delta_{A^{\mathbf{b}}}) = \Big\{(2j+1)\mathbf{b} - j(j+1)\,;\ j \in \mathbb{N}\,,\ j < \mathbf{b} - \frac{1}{2}\Big\}, \quad \Big(\text{if } \frac{1}{2} < \mathbf{b}\Big).$$

3.3.3. Non-Weyl-type asymptotics (high energy). Let us assume that

$$A_j(x,y) \in C^2(\mathbb{H};\mathbb{R}), \quad \forall j. \tag{3.12}$$

It is well known that $H_1(A)$ defined by (3.7) is then essentially self-adjoint on $L^2(\mathbb{H})$, see for example [54]. We assume moreover the following magnetic bottles-type assumptions.

- $$\mathbf{b}(x,y) \to +\infty \quad \text{as} \quad d(x,y) \to +\infty, \tag{3.13}$$
  ($d(x,y)$ denotes the hyperbolic distance from (x,y) to the point $(0,1)$).
- $\exists\, C_0 > 0$ such that, for any vector field X on $\mathbb{H}$,
  $$|X\widetilde{\mathbf{b}}| \le C_0(|\widetilde{\mathbf{b}}|+1)\sqrt{g(X,X)}\,; \tag{3.14}$$

Theorem 3.14. *Under the assumptions* (3.12), (3.13) *and* (3.14)

1) *the operator* $H_1(A)$ *has a compact resolvent.*
2) *for any* $\delta \in]\frac{1}{3},\frac{2}{5}[$, *there exists a constant* $C > 0$ *such that*

$$\frac{1}{2\pi}\int_{\mathbb{H}}\left(1-\frac{C}{(\mathbf{b}(m)+1)^{(2-5\delta)/2}}\right)\mathbf{b}(m)\sum_{k=0}^{+\infty}\left[\lambda(1-C\lambda^{-3\delta+1})-\frac{1}{4}-(2k+1)\mathbf{b}(m)\right]_+^0 dv$$
$$\le N(\lambda, H_1(A)) \le \tag{3.15}$$
$$\frac{1}{2\pi}\int_{\mathbb{H}}\left(1+\frac{C}{(\mathbf{b}(m)+1)^{(2-5\delta)/2}}\right)\mathbf{b}(m)\sum_{k=0}^{+\infty}\left[\lambda(1+C\lambda^{-3\delta+1})-\frac{1}{4}-(2k+1)\mathbf{b}(m)\right]_+^0 dv$$

Remark 3.15. Comparing this result with the one obtained in [6] and in particular with the formula (3.6), it turns out that they differ only by the additional term $-\frac{1}{4}$, which comes from the geometry of the problem. This term becomes really significant in the following corollary:

Corollary 3.16. *Under the assumptions of Theorem* 3.14 *and if the function*

$$\omega(\mu) = \int_{\mathbb{H}} [\mu - \mathbf{b}(m)]_+^0\, dv$$

verifies

$$\exists\, C_1 > 0 \text{ s.t. } \forall\, \mu > C_1,\ \forall\, \tau \in]0,1[,\quad \omega\left((1+\tau)\,\mu\right)-\omega(\mu) \le C_1\, \tau\, \omega(\mu), \tag{3.16}$$

then

$$N(\lambda; H_1(A)) \sim \frac{1}{2\pi}\int_{\mathbb{H}} \mathbf{b}(m)\sum_{k\in\mathbb{N}}\left[\lambda-\frac{1}{4}-(2k+1)\mathbf{b}(m)\right]_+^0 dv. \tag{3.17}$$

The assumption (3.16) is satisfied for example when $\omega(\lambda) \sim \alpha\lambda^k \ln^j(\lambda)$ when $\lambda \to +\infty$, with $k > 0$, or $k = 0$ and $j > 0$.

For example this allows us to consider magnetic fields of the type

$$\mathbf{b}(x,y) = \left(\frac{x}{y}\right)^{2j} + g(y),$$

with $j \in \mathbb{N}^\star$ and $g(y) = p_1(y) + p_2(1/y)$, where $p_1(s)$ and $p_2(s)$ are, for large s, polynomial functions of order ≥ 1.

The function $\omega(\lambda)$ indeed verifies in that case $\omega(\lambda) \sim \alpha\lambda^{\frac{1}{2j}}\ln(\lambda)$ when $\lambda \to +\infty$. and

$$N(\lambda; H_1(A)) \sim \frac{C}{2\pi}\lambda^{1+1/2j}\ln(\lambda) .$$

In next section we give an outline of the proof of theorem 3.14, by describing the techniques specific to the hyperbolic context: definition of a diffeomorphism from $\mathbb{R}^2$ to $\mathbb{H}$, control of the magnetic field by a constant one in a suitable rectangle, partition of $\mathbb{R}^2$ into such appropriate rectangles, so that we can apply the min-max variational method.

3.3.4. Outline of the proof.

A diffeomorphism from $\mathbb{R}^2$ *to* $\mathbb{H}$. Let us consider the diffeomorphism

$$\phi : \mathbb{R}^2 \to \mathbb{H}$$

$(x, y) = \phi(x, t) := (x, e^t)$ which induces a unitary operator

$$\widehat{U} : L^2(\mathbb{H}; dv) \to L^2(\mathbb{R}^2; dxdt)$$

$(\widehat{U}f)(x, t) = e^{-t/2}f(x, e^t)$ for any $f \in L^2(\mathbb{H})$.

The quadratic form associated to $H_1(A)$ is given by $(\forall u \in L^2(\mathbb{H}))$

$$q(u) = \int_{\mathbb{H}} \left[|y(D_x - A_1)u|^2 + |y(D_y - A_2)u|^2\right] \frac{dxdy}{y^2}.$$

Writing $\tilde{A}_i(x, t) = A_i(x, e^t), i = 1, 2$, and $w = \widehat{U}u$, we get after computation

$$q(u) = \widehat{q}^{\tilde{A}}(w) = \int_{\mathbb{R}^2} \left[|e^t(D_x - \tilde{A}_1)w|^2 + |(e^{-t/2}D_t e^{t/2} - e^t\tilde{A}_2)w|^2\right] dxdt .$$

The operator associated to $\widehat{q}^{\tilde{A}}$ is $\widehat{H}(\tilde{A}) = \widehat{U}H_1(A)\widehat{U}^{-1}$.

This gives $\widehat{H}(\tilde{A})e^{2t}(D_x - \tilde{A}_1)^2 + (D_t - e^t\tilde{A}_2)^2 + 1/4$.

The additional term $1/4$ appears here naturally as a by-product of the transformation which allows us to deal with a problem in $\mathbb{R}^2$ instead of the initial problem in $\mathbb{H}$.

Gauge. We want to work with a gauge such that $A_2 = 0$. Since

$$\widetilde{\mathbf{b}} = y^2(\partial_x A_2 - \partial_y A_1)$$

one can take

$$A_1(x, y) = -\int_1^y \frac{\widetilde{\mathbf{b}}(x, s)}{s^2} ds$$

The associated quadratic form is

$$\widehat{q}^{\tilde{A}}(w) = \int_{\mathbb{R}^2} \left[|e^t(D_x - \tilde{A}_1)w|^2 + |D_t w|^2 + 1/4|w|^2\right] dxdt .$$

Localization. According to the assumption (3.14) we can control the magnetic field by a constant one on an appropriate rectangle:

We set

$$\Omega(x_0, y_0, a, \varepsilon_0) := \{(x, y) \ / \ |x - x_0| \leq a\varepsilon_0 \ y_0, \ |y - y_0| \leq \varepsilon_0 y_0\}$$

($a > 0$ and $\varepsilon_0 > 0$ small enough).

Lemma 3.17. *There exists $C_1 > 0$ such that, for any $(x_0, y_0) \in \mathbb{H}$ with $\mathbf{b}(x_0, y_0) > 1$, the following holds*

$$\frac{1}{C_1} \ \mathbf{b}(x_0, y_0) \leq \mathbf{b}(x, y) \leq C_1 \ \mathbf{b}(x_0, y_0) \quad \forall \ (x, y) \in \Omega(x_0, y_0, a, \varepsilon_0).$$

Partition of $\mathbb{R}^2$. For any $\alpha \in \mathbb{Z}^2$, let us denote by $K(\alpha)$ the rectangle

$$K(\alpha) \ = \ \left]-\frac{e^{\alpha_2}}{2} + e^{\alpha_2}\alpha_1 \ , \ e^{\alpha_2}\alpha_1 + \frac{e^{\alpha_2}}{2}\right[\ \times \ \left]-\frac{1}{2} + \alpha_2 \ , \ \alpha_2 + \frac{1}{2}\right[. \qquad (3.18)$$

Therefore $\mathbb{R}^2 = \cup_\alpha \overline{K}(\alpha)$ and $K(\alpha) \cap K(\beta) = \emptyset$ for any $\alpha \neq \beta$. According to Lemma 3.17, it is possible to subdivide each $K(\alpha)$, (if necessary), into $M(\alpha)$ rectangles:

$$\overline{K}(\alpha) \ = \ \cup_{j=1}^{M(\alpha)} \overline{K_{\alpha,j}} \qquad (3.19)$$

$$K_{\alpha,j} \ = \ \left]-\frac{\epsilon_{\alpha,j} e^{t_{\alpha,j}}}{2} + x_{\alpha,j} \ , \ x_{\alpha,j} + \frac{\epsilon_{\alpha,j} e^{t_{\alpha,j}}}{2}\right[\ \times \ \left]-\frac{\epsilon_{\alpha,j}}{2} + t_{\alpha,j} \ , \ t_{\alpha,j} + \frac{\epsilon_{\alpha,j}}{2}\right[,$$

with

$$\frac{1}{a_0(1 + \mathbf{b}^{\delta_0}(x_{\alpha,j}, e^{t_{\alpha,j}}) \ )} \ \leq \ \epsilon_{\alpha,j} \ \leq \ \frac{a_0}{(1 + \mathbf{b}^{\delta_0}(x_{\alpha,j}, e^{t_{\alpha,j}}) \ )}, \qquad (3.20)$$

and such that $K_{\alpha,k} \cap K_{\alpha,j} \ = \ \emptyset$ if $k \neq j$.

This lemma is the hyperbolic version of Lemma 3.11 for the Euclidean case.

The partition $\mathbb{R}^2 \ = \ \cup_\alpha \overline{K}(\alpha)$ and the partition on $\mathbb{H}$ obtained after applying the diffeomorphism ϕ are represented on Figures 2 and 3.

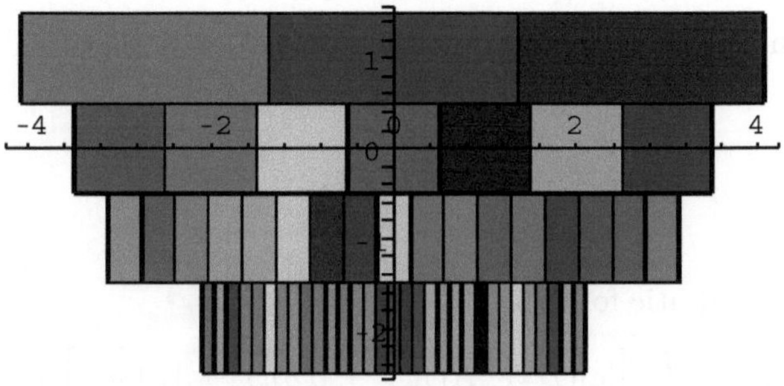

FIGURE 2. Partition of $\mathbb{R}^2$ by the rectangles $\overline{K}(\alpha)$.

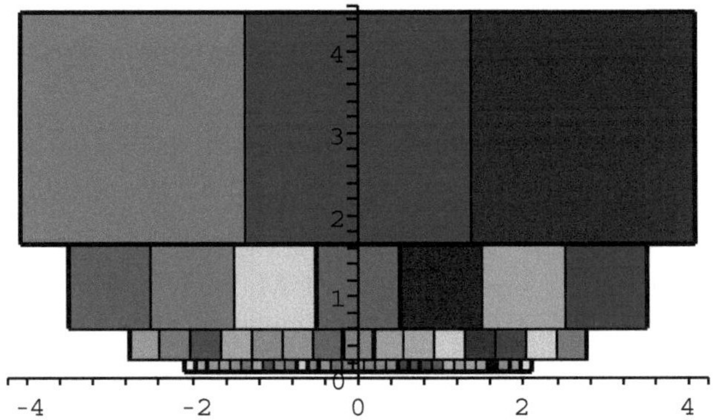

FIGURE 3. Partition of $\mathbb{H}$ by the cubes $\phi(\overline{K}(\alpha))$.

3.4. Geometrically finite hyperbolic surfaces

3.4.1. Introduction. Concerning magnetic bottles in hyperbolic geometry the min-max method can be generalized to the geometrically finite hyperbolic surfaces, in the case when these manifolds are of infinite volume. Such manifolds contain cusps and funnels [43].

Actually, when the hyperbolic manifolds are compact the result is given by [7] in the more general context of compact Riemannian manifolds, where it is shown that the Weyl asymptotics hold. For the case of non compact manifolds of finite volume we refer to [20], where the authors study examples for which the Weyl formula is still valid:

$$N(\lambda) \sim_{+\infty} \frac{\lambda}{4\pi}|\mathbf{M}|.$$

It seems to be the standard result in this context.

In the case of the Poincaré half-plane, $\mathbf{M} = \mathbb{H}$, we have seen previously that the Weyl formula does not hold:

$$\lim_{\lambda \to +\infty} \lambda^{-1} N(\lambda) = +\infty.$$

For example when $\mathbf{b}(z) = a_0^2(x/y)^{2m_0} + a_1^2 y^{m_1} + a_2^2/y^{m_2}$, $a_j > 0$ and $m_j \in \mathbb{N}^\star$, then

$$N(\lambda) \sim_{+\infty} \lambda^{1+1/(2m_0)} \ln(\lambda)\alpha(m_0, m_1, m_2) \, .$$

It turns out [43] that it is still the case when $\mathbf{M}$ has an infinite area and is geometrically finite, and if we adapt the preceding example to this new situation, i.e., m_0 is absent, m_1 appears in the cusps and m_2 in the funnels, we get

$$N(\lambda) \sim_{+\infty} \lambda^{1+1/m_2}\alpha(m_2) \, .$$

The interesting point is that the cusps do not contribute to the leading part of $N(\lambda)$.

Let us explain the result, and first what is such a surface.

3.4.2. Definition. If $(\mathbf{M},\ g)$ is a smooth connected Riemannian manifold of dimension two, it is called a geometrically finite hyperbolic surface of infinite area if it can be decomposed in the following way:

$$\mathbf{M} \;=\; \left(\bigcup_{j=0}^{J_1} M_j \right) \bigcup \left(\bigcup_{k=1}^{J_2} F_k \right) \;; \tag{3.21}$$

where the M_j and the F_k are open sets of $\mathbf{M}$, such that the closure of M_0 is compact, and if $J_1 > 0$, the other M_j are cuspidal ends of $\mathbf{M}$, and the F_k are funnel ends of $\mathbf{M}$.

This means that, for any j, $1 \leq j \leq J_1$, there exist strictly positive constants a_j and L_j such that M_j is isometric to $\mathbb{S}\times]a_j^2, +\infty[$, equipped with the metric

$$ds_j^2 \;=\; y^{-2}(\; L_j^2 \; d\theta^2 \;+\; dy^2 \;)\;. \tag{3.22}$$

($\mathbb{S} \;=\; \mathbb{S}^1$ is the unit circle.)

In the same way, for any k, $1 \leq k \leq J_2$, there exist strictly positive constants α_k and τ_k such that F_k is isometric to $\mathbb{S}\times]\alpha_k^2, +\infty[$, equipped with the metric

$$ds_k^2 \;=\; \tau_k^2 \cosh^2(t) d\theta^2 \;+\; dt^2\;. \tag{3.23}$$

Moreover, for any two integers $j, k > 0$, we have

$$M_j \cap F_k = \emptyset \quad \text{and} \quad M_j \cap M_k = F_j \cap F_k = \emptyset \quad \text{if} \;\; j \neq k.$$

3.4.3. Assumptions on the magnetic field. Let us choose some $z_0 \in M_0$ and let us define

$$d \;:\; \mathbf{M} \;\rightarrow\; \mathbb{R}_+\;; \quad d(z) \;=\; d_g(z, z_0)\;; \tag{3.24}$$

$d_g(\,.\,,\,.\,)$ denotes the distance with respect to the metric g.

We assume the smooth one-form A to be given such that the magnetic field $\widetilde{\mathbf{b}}$ satisfies

$$\lim_{d(z)\to\infty} \;\; \mathbf{b}(z) \;=\; +\infty\;. \tag{3.25}$$

If $J_1 > 0$, there exists a constant $C_1 > 0$ such

$$|X\widetilde{\mathbf{b}}(z)| \;\leq\; C_1(\mathbf{b}(z) \;+\; 1)e^{d(z)}|X|_g\;; \tag{3.26}$$

$\forall z \in M_j$, $\forall X \in T_z\mathbf{M}$ and $\forall j = 1, \ldots, J_1$.

There exists a constant $C_2 > 0$ such

$$|X\widetilde{\mathbf{b}}(z)| \;\leq\; C_2(\mathbf{b}(z) \;+\; 1)|X|_g\;; \tag{3.27}$$

$\forall z \in F_k$, $\forall X \in T_z\mathbf{M}$ and $\forall k = 1, \ldots, J_2$.

3.4.4. Asymptotics for large energies.

Theorem 3.18. *Under the above assumptions, $-\Delta_A$ has a compact resolvent and for any $\delta \in \left]\frac{1}{3}, \frac{2}{5}\right[$, there exists a constant $C > 0$ such that*

$$\frac{1}{2\pi}\int_{\mathbf{M}}\left(1 - \frac{C}{(\mathbf{b}(m)+1)^{(2-5\delta)/2}}\right)\mathcal{N}(\lambda(1 - C\lambda^{-3\delta+1}) - \frac{1}{4}, \mathbf{b}(m))\, dm$$
$$\leq N(\lambda, -\Delta_A) \leq \tag{3.28}$$
$$\frac{1}{2\pi}\int_{\mathbf{M}}\left(1 + \frac{C}{(\mathbf{b}(m)+1)^{(2-5\delta)/2}}\right)\mathcal{N}(\lambda(1 + C\lambda^{-3\delta+1}) - \frac{1}{4}, \mathbf{b}(m))\, dm$$

where

$$\mathcal{N}(\mu, \mathbf{b}(m)) = \mathbf{b}(m)\sum_{k=0}^{+\infty}[\mu - (2k+1)\mathbf{b}(m)]_+^0 \quad \textit{if } \mathbf{b}(m) > 0\,,$$

and

$$\mathcal{N}(\mu, \mathbf{b}(m)) = \mu/2 \quad \textit{if } \mathbf{b}(m) = 0\,.$$

$[\rho]_+^0$ is the Heaviside function:

$$[\rho]_+^0 = \begin{cases} 1\,, & \textit{if} \quad \rho > 0 \\ 0\,, & \textit{if} \quad \rho \leq 0\,. \end{cases}$$

The Theorem remains true if we replace $\displaystyle\int_{\mathbf{M}}$ by $\displaystyle\sum_{k=1}^{J_2}\int_{F_k}$, due to the fact that the other parts are bounded by $C\lambda$.

Corollary 3.19. *Under the assumptions of Theorem* 3.18 *and if the function*

$$\omega(\mu) = \int_{\mathbf{M}}[\mu - \mathbf{b}(m)]_+^0\, dm$$

satisfies, $\exists\, C_1 > 0$ s.t. $\forall\, \mu > C_1\,,\ \forall\, \tau \in]0,1[$,

$$\omega\left((1+\tau)\,\mu\right) - \omega(\mu) \leq C_1\,\tau\,\omega(\mu)\,, \tag{3.29}$$

then

$$N(\lambda; -\Delta_A) \sim \frac{1}{2\pi}\int_{\mathbf{M}}\mathcal{N}\left(\lambda - \frac{1}{4}, \mathbf{b}(m)\right)dm\,. \tag{3.30}$$

For example this allows us to consider magnetic fields of the following type:

$$\text{on } F_k, \quad \mathbf{b}(\theta, t) = p_k(1/\cosh(t)),$$
$$\text{and on } M_j,\ j > 0, \quad \mathbf{b}(\theta, y) = q_j(y)\,,$$

where the $p_k(s)$ and the $q_j(s)$ are, for large s, polynomial functions of order ≥ 1. In this case, if d is the largest order of the $p_k(s)$, then

$$N(\lambda; -\Delta_A) \sim \alpha\lambda^{1+1/d}\,,$$

for some constant $\alpha > 0$, depending only on the funnels F_k where the order of $p_k(s)$ is d.

4. A Neumann problem with magnetic field

4.1. A problem arising from super-conductivity

In super-conductivity theory the following question has to be answered: can we minimize, for a given open set Ω of $\mathbb{R}^d$, and a given potential $A \in H^1(\Omega;\mathbb{R}^d)$, the Ginzburg-Landau functional:

$$\mathcal{G}(\psi,\tilde{A}) = \int_\Omega |(\nabla - i\kappa\tilde{A})\psi|^2 + \frac{\kappa^2}{2}(|\psi|^2-1)^2\, dx \cdots$$
$$\cdots + \kappa^2 \int_\Omega |\overrightarrow{\mathrm{rot}}\tilde{A} - \sigma\overrightarrow{\mathrm{rot}}A|^2\, dx$$

on the set $(\psi,\tilde{A}) \in H^1(\Omega;\mathbb{C}) \times H^1(\Omega;\mathbb{R}^d)$. κ and σ denote parameters related to the intensity of the magnetic field, and the dimension d considered is either 2 or 3. $(0,\sigma A)$ turns out to be a trivial critical point of $\mathcal{G}(\psi,\tilde{A})$; in order to study the Hessian matrix of the functional at this point we are then reduced to investigate the spectral properties, (modulo the parameters κ and σ) of the magnetic Laplacian $H_h = ((h\nabla - iA))^2$, with the Neumann-type boundary conditions:

$$\nu(x)\cdot(h\nabla - iA)u(x) = 0 \quad \forall x \in \delta\Omega\ .$$

A lot has been done to understand the properties of this operator. In particular it is known from the works by K. Lu and X.B. Pan [34]–[36] and by B. Helffer and A. Morame [22]–[24] that unlike the Dirichlet case, the lower bound of its spectrum can be less than $hb = h\inf_\Omega \|B\|$ if $B = dA$ denotes the magnetic field. This comes from the following fundamental fact, which makes the basic difference between Dirichlet and Neumann problems:

If we consider the Neumann operator on $L^2(\mathbb{R}_+)$ defined by

$$Q_x = D_t^2 + (t-x)^2,$$

and if we denote by $\mu(x)$ its first eigenvalue, then

$$\inf_{x\in\mathbb{R}} \mu(x) = \mu(x_0) = \Theta_0 < 1\ .$$

If we consider the Dirichlet operator the corresponding quantity is equal to 1.

As a consequence, in the case where $b \geq \Theta_0\, b'$, (with $b' = \inf_{\delta\Omega}\|B\|$), we get that the lower bound of the spectrum for the Neumann problem is $h\Theta_0\, b'$.

Furthermore, for a constant non zero B, any normalized fundamental eigenfunction is localized exponentially (for h going to 0) in the neighbourhood of points of the boundary with maximal curvature (see [22] for dimension 2 and [23] for dimension 3). Superconductivity comes precisely from this crucial feature. We should also mention the papers by B. Helffer and S. Fournais, in particular [16] for asymptotic estimates of low eigenvalues in dimension 2, and [17] about the "critical fields", which are responsible for the transitions from superconducting states to normal states. This list of papers on the subject is far from being exhaustive...

4.2. The spectrum in the case of the half-space, for a constant field and for $h = 1$

Let us consider, for (t, x, y) in $\Omega = \mathbb{R}_+ \times \mathbb{R}^2$, the Neumann realization of the magnetic Laplacian

$$H = (D_t - A_1)^2 + (D_x - A_2)^2 + (D_y - A_3)^2$$

where $D_s = -i\left(\frac{\partial}{\partial s}\right)$.

Let us denote by b the norm of B, and by θ the angle between $B = dA$, seen as a three-dimensional vector field, and the boundary $\partial\Omega$.

This implies that a suitable choice for the gauge A is the 1-form

$$A = b(x \sin\theta - t\cos\theta)dy$$

so that the operator H can be written as

$$H_\theta^b = D_t^2 + D_x^2 + (D_y - b(x\sin\theta - t\cos\theta))^2 \ .$$

By homogeneity we get:

$$\sigma(H_\theta^b) = b\sigma(H_\theta)$$

with

$$H_\theta = D_t^2 + D_x^2 + (D_y - (x\sin\theta - t\cos\theta))^2 \ .$$

- $\boldsymbol{\theta = 0}$. The spectrum of the Neumann operator H_0 is absolutely continuous. More precisely one has:

$$\sigma(H_0) = \sigma_{\mathrm{ac}}(H_0) = [b\Theta_0, +\infty[\ . \tag{4.1}$$

- $\boldsymbol{\theta = \frac{\pi}{2}}$. The spectrum of $H_{\frac{\pi}{2}}$ is still absolutely continuous but

$$\sigma(H_{\frac{\pi}{2}}) = \sigma_{\mathrm{ac}}[b, +\infty[\ . \tag{4.2}$$

- $\boldsymbol{\theta \in \left]0, \frac{\pi}{2}\right[}$. The spectrum of H_θ is no longer absolutely continuous as proved by K. Lu and X-B. Pan [35], (see also [23]).

We prove in [41] the following:

Theorem 4.1. *If $\theta \in \left]0, \frac{\pi}{2}\right[$,*

$$\sigma(H_\theta^b) \cap \left]-\infty, b\right[= \{b\nu_1(\theta),\ b\nu_2(\theta), \dots,\ b\nu_j(\theta),\ b\nu_{j+1}(\theta), \dots\}\ . \tag{4.3}$$

(Each $b\nu_j(\theta)$ is an eigenvalue of infinite multiplicity of H_θ).

To prove this result we first observe that $\sigma(H_\theta) = \bigcup_{\tau\in\mathbb{R}} \sigma(H_{\theta,\tau})$, where $H_{\theta,\tau}$ denotes the Neumann realization in the half-plane $F = \mathbb{R}_+ \times \mathbb{R}$ of the operator

$$H_{\theta,\tau} D_t^2 + D_x^2 + (\tau - (x\sin\theta - t\cos\theta))^2.$$

Furthermore using for any τ the change of coordinates $x \to x - \dfrac{\tau}{b\sin\theta}$, we see that $\sigma(H_{\theta,\tau}) = \sigma(P_\theta)$, with

$$P_\theta = D_t^2 + D_x^2 + (t\cos\theta - x\sin\theta)^2\ ,$$

and thus the spectrum of H_θ is essential and given by:

$$\sigma(H_\theta) = \sigma_{\text{ess}}(H_\theta) = \sigma(P_\theta). \tag{4.4}$$

In [35], (see also [23]), it was proved that

$$\inf\ \sigma(P_\theta) = \nu(\theta) < 1 = \inf\ \sigma_{\text{ess}}(P_\theta), \tag{4.5}$$

so there exists a countable set of eigenvalues $(\nu_j(\theta))_{j\in I}$, $(I \subset \mathbb{N})$, contained in $[\nu(\theta), 1[$.

4.3. Non-Weyl-type asymptotics when the field is nearly tangent to the boundary

For any $d \leq 1$ let us denote by $N(d,\ P_\theta)$ the number of eigenvalues of P_θ in $\left]-\infty, d\right[$:

$$N(d,\ P_\theta) = \text{Tr}(E_{]-\infty,d[}(P_\theta)) = \sharp\{j;\ \nu_j(\theta) < d\}\ . \tag{4.6}$$

We prove the following results [41]:

Theorem 4.2. *For any* $\theta\ \in \left]0, \frac{\pi}{2}\right[$, P_θ *admits a finite number of eigenvalues in* $\left]-\infty, 1\right[$, *and there exists a constant* $C \geq 1$ *such that*

$$N(1,\ P_\theta) \leq \frac{C}{\sin\theta}\ . \tag{4.7}$$

It can be noticed that the upper bound of $N(1,\ P_\theta)$ goes to infinity when the angle θ between the magnetic field and the boundary tends to zero. Therefore we can consider θ (or more precisely $\dfrac{\sin\theta}{\sqrt{\cos\theta}}$) as a semi-classical parameter, and using once more min-max techniques we give a non-Weyl-type asymptotic estimate of $N(d, P_\theta)$ for $d < 1$ [41]:

Theorem 4.3. *If* $d \in \left]\Theta_0,\ 1\right[$, *there exists a constant* $C_d > 0$ *such that*

$$\left| N(d, P_\theta)\ -\ \frac{1}{2\pi\sin\theta}\ \int_{\mathbb{R}} \left[d - \mu(x)\right]_+^{1/2} dx \right| \leq C_d\ . \tag{4.8}$$

In this expression clearly appears an "effective" potential, as in the previous case of degenerate potentials. This is not so surprising, since the operator we finally study is

$$P_\theta = D_t^2 + D_x^2 + (t\cos\theta - x\sin\theta)^2\ ,$$

which turns out to be a Schrödinger operator with the degenerate potential $V_\theta(x,t) = (t\cos\theta - x\sin\theta)^2$. The "effective" potential here is the function of one variable $\mu(x)$ previously introduced, which is responsible for the superconductivity.

5. A problem of magnetic bottle in classical mechanics

5.1. The Lorentz equation

The motion in $\mathbb{R}^3$ of a particle of mass m and charge e in a magnetic field $\vec{B}$ can be described by Lorentz equation:

$$m\ddot{x} = e\dot{x} \wedge \vec{B} \, .$$

To this equation corresponds the following Lagrangian (for $m = e = 1$):

$$\mathcal{L}(x, \dot{x}) = \frac{1}{2} \, \dot{x}^2 + \dot{x} \cdot A(x) \, , \tag{5.1}$$

where A denotes a magnetic potential: $\overrightarrow{rot} A = \vec{B}$. To obtain the associated Hamiltonian, we compute the conjugate momenta

$$\xi_j = \frac{\partial \mathcal{L}}{\partial \dot{x}_j} \tag{5.2}$$

which writes $\xi = \dot{x} + A(x)$ so we get:

$$\mathcal{H}(x, \xi) = \xi \dot{x} - \mathcal{L}(x, \dot{x}) \;=\; \frac{1}{2}(\xi - A(x))^2 \, . \tag{5.3}$$

When $\vec{B}$ is a constant field (in time and position) the Hamiltonian is integrable, the trajectories are helicoidal and the axis of the motion is the direction of the field. There are three numbers conserved during the motion (the integrals of motion) which are the energy (e.g., the Hamiltonian itself), the Larmor radius $\rho = \frac{v_\perp}{B}$ and the magnetic moment $I = \frac{v_\perp^2}{2B}$. We denote respectively by $v_\perp$ the orthogonal component (to field lines) of the velocity and by B the norm of $\vec{B}$.

Remark 5.1. Applying Weyl quantification to $\mathcal{H}(x, \xi)$ we obtain the Schrödinger operator defined on $L^2(\mathbb{R}^3)$ by

$$H_h(A) = \sum_{j=1}^{3} (\frac{h}{i} \frac{\partial}{\partial x_j} - A(x_j))^2 \, .$$

In the case when $\vec{B}$ is a constant vector field the spectrum of the operator $H_h(A)$ is composed of eigenvalues of infinite multiplicity, which are the Landau levels $\lambda_j(h) = h \, (2j+1) \, B$.

5.2. Adiabatic invariants

Let us consider a magnetic field slowly varying in position so that it is almost constant throughout a whole rotation of the particle: the motion is approximatively a circle and the center of this circle (the guiding center) is slowly moving along the direction of the field, with a very small rotation period.

Under the previous conditions the Hamiltonian slowly depends on the position variables, except for one (denoted by x_0); it can be written as

$$\mathcal{H} = \mathcal{H}(x, \epsilon x_0, \xi, \xi_0),$$

where ϵ is a small parameter.

Let us assume that, for the Hamiltonian $\mathcal{H}_0(x,\xi)$ obtained by fixing the value of ϵx_0 and ξ_0, there exist closed trajectories in phase space (with a non vanishing frequency). Then one can introduce the action-angle variables (I,ϕ). The action variable $I((x,\epsilon x_0,\xi,\xi_0)$ corresponds to the magnetic moment; it is not an integral of motion as it is in the constant case, but it turns out from the method of moyennisation [1] that it is an adiabatic invariant, which means more precisely:

$$\exists c>0 \ t.q. \ |I((x(t),\epsilon x_0(t),\xi(t),\xi_0(t)) - I((x(0),\epsilon x_0(0),\xi(0),\xi_0(0))| < \ c\,\epsilon$$

for $0 \le t \le 1/\epsilon$.

Performing symplectic transformations one can get the invariance of I to all orders ([32], [46]). Furthermore, if the magnetic field is a convex function along the field lines (seen as a function of the arc length s) there exists another invariant, which is longitudinal and given by $J = \oint \frac{v_\parallel^2}{B} ds$. The trajectory is actually reflected at the points s_1, s_2 verifying $IB(s_i) = \mathcal{H}$, and the integral is computed on a whole oscillation [48].

M. Gardner [18] proves the invariance of this quantity to all orders.

V.I. Arnold [2] considers the following question: is it possible to get the particles not only adiabatically but really confined? Considering a magnetic field with symmetry axis he wrote the corresponding Hamiltonian, with two degrees of freedom, as a perturbation of an integrable one, for which the motion in phase space is on a torus.

Under a non-degeneracy condition (the ratio of frequencies varies in time) he applies KAM theorem to get that the invariant tori are not all destroyed under the perturbation and he concludes by a dimension argument that the action I is a perpetually adiabatic invariant.

The difficult point here is to check the non-degeneracy condition, according to the fact that the ratio of frequencies is of small order. Arnold checks that condition only in the special case when the Hamiltonian writes $\mathcal{H} = \frac{1}{2}(\xi_1^2 + \xi_0^2) + U(x_1, x_0)$ with: $U(x_1,x_0) = \frac{1}{2}x_1^2(1+\epsilon^2 x_0^2)$.

Another method consists in applying a theorem of J. Moser [44] which gives the existence of periodic solutions for systems next to an integrable one. M. Braun [5] proves by this way the existence of a region where the particles submitted to the earth magnetic field are indefinitely retained.

In [58] we apply Moser's theorem in the case of a magnetic field which is linear and symmetric in position

$$\vec{B}(x,y,z) = (x,y,-2z)\,, \tag{5.4}$$

and we get an open set of initial conditions for which the motion remains bounded.

This field is actually of the"magnetic bottle" type, since its norm tends to infinity as position tends to infinity. As already mentioned, the operator obtained from the Hamiltonian by Weyl's quantification has a compact resolvent. The result of quantum mechanics seems to be stronger compared with the classical one. This is a well-known fact that it is harder to get informations on the classical motion than on the spectrum of the quantum operator ...

5.3. Bounded trajectories

5.3.1. The Hamiltonian in cylindrical coordinates. According to the symmetry of the magnetic field (5.4), we introduce cylindrical coordinates. Considering B as a 2-form we have

$$B = d(-r^2 z) \wedge d\theta,$$

so we can choose the following gauge

$$A = (A_r, A_\theta, A_z) = (0, -rz, 0).$$

The field lines are characterized by $\theta =$ constant and $r^2 z$ =constant.

The conjugate moments, (defined by (5.2)) have the following expression

$$(\xi_r, \xi_\theta, \xi_z) = (\dot{r}, r^2(\dot{\theta} - z), \dot{z}) \ ,$$

so we get, using formula (5.3)

$$\mathcal{H}(r, \theta, z, \xi_r, \xi_\theta, \xi_z) = \frac{1}{2}(\xi_r^2 + \xi_z^2) + \frac{1}{2r^2}(\xi_\theta + r^2 z)^2 \ . \tag{5.5}$$

Writing the second equation of Hamilton

$$\dot{\xi_\theta} = -\frac{\partial \mathcal{H}}{\partial \theta} \tag{5.6}$$

one gets the existence of an integral of motion, which is ξ_θ.

Dimension has been reduced according to the symmetry of the problem. We have to study a Hamiltonian with 2 degrees of freedom which is defined as follows:

$$\mathcal{H}_M(r, z, \xi_r, \xi_z) = \frac{1}{2}(\xi_r^2 + \xi_z^2) + \frac{(M + r^2 z)^2}{2r^2} \ . \tag{5.7}$$

5.3.2. The reduced Hamiltonian and the magnetic field lines. The value of $M = \xi_\theta$ is fixed by the initial conditions, so is the energy E (E is the value taken by the Hamiltonian, and it is a constant of motion too). M may be negative. let us consider the magnetic field line (L_M) defined by $r^2 z = -M$. There exists a unique point Ω_M such that the norm B at that point is minimal on (L_M). Thus, to any point $P(r, z)$ we associate new coordinates (u, v) as follows:

- v is the distance from $P(r, z)$ to the magnetic field line (L_M)
- u is the arc length between the projection of P on (L_M) to Ω_M.

We denote by $k(u)$ the curvature of (L_M) at the point $(u, 0)$ and we set:

$$\mathcal{H}_0(u, v, \xi_u, \xi_v) = \frac{1}{2}\left(\frac{\xi_u^2}{[1 + vk(u)]^2} + \xi_v^2 + B^2(u, 0)v^2\right) . \tag{5.8}$$

According to the new coordinates the Hamiltonian $\mathcal{H}_M(u, v, \xi_u, \xi_v)$ obtained from the expression (5.7) can be reexpressed as a perturbation of $\mathcal{H}_0(u, v, \xi_u, \xi_v)$ in the following way:

Proposition 5.2. *For any initial conditions satisfying*

$$E < 2|M|^{4/3} \quad \text{and} \quad E < \frac{\epsilon^2}{16}|M|^{2/3} \quad \text{(CI)}$$

- *the distance v from the trajectory to the magnetic field line (L_M) is less than ϵ,*
- *there exists a constant C (depending only on E and M) such that*

$$|\mathcal{H}_M(u, v, \xi_u, \xi_v) - \mathcal{H}_0(u, v, \xi_u, \xi_v)| < C\epsilon^3.$$

Remarks

1) The first condition is a consequence of the second one for ϵ small enough.
2) The proposition comes from the fact that according to the inequality $\frac{(M+r^2z)^2}{2r^2} < E$ the motion has to remain inside a strip $\mathcal{B}$ around (L_M) (see Figure 4).

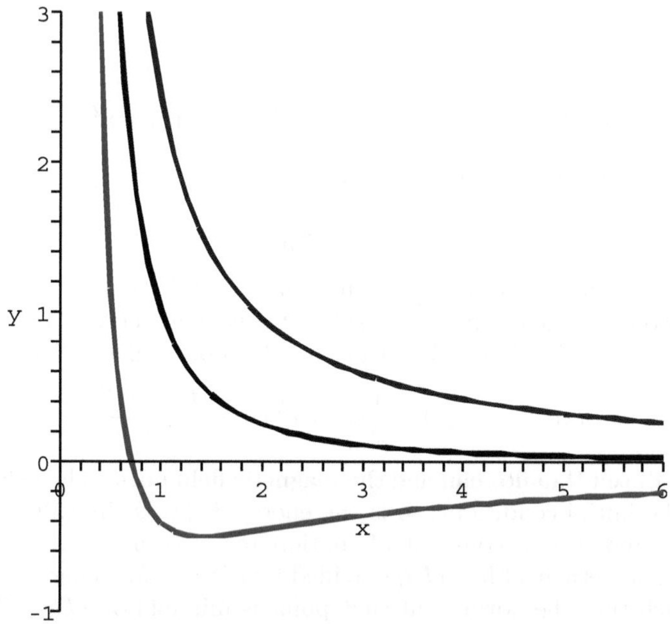

FIGURE 4. The strip $\mathcal{B}$ and the magnetic field line (L_M).

3) Since v is small the original Hamiltonian can be seen indeed as a perturbation of the Hamiltonian $\frac{1}{2}\left(\xi_u^2 + \xi_v^2 + B^2(u,0)v^2\right)$ which corresponds quantically to an operator with a degenerate potential, precisely of the form described in section 2.2.
4) The Hamiltonian $\mathcal{H}_u = \frac{1}{2}\left(\xi_v^2 + B^2(u,0)v^2\right)$ represents for a fixed value of u the energy of a harmonic oscillator. The point (v, ξ_v) moves along the ellipse $\mathcal{H}_u$ = constant for some constant depending on u and the corresponding

action, which is (up to a factor 2π) the area enclosed by the ellipse, has the following expression

$$I_u = \frac{\xi_v^2 + B^2(u,0)v^2}{2\,B(u,0)} \,.$$

5) Conditions (CI) entail the following inequality

$$B(u,0)I_u < \frac{\epsilon^2}{16}|M|^{2/3} + C\epsilon^3 \,.$$

As a consequence, if I_u is bounded from below by a constant independent of time, we get an upper bound for $B(u,0)$ and hence for the trajectory itself since $B(u,0)$ is an increasing function of u.

5.3.3. Action variables. We set

$u = \epsilon u_1,\ v = \epsilon v_1,\ \xi_u = \epsilon\ \xi_{u_1},\ \xi_v = \epsilon\ \xi_{v_1},\ \mathcal{H}_M = \epsilon^2\ \mathcal{K}_M,\ E = \epsilon^2 E'$.

We perform some symplectic transformations (in the language of mechanics, we perform some changes of canonical variables) in order to get explicit action-angle coordinates (I, J, ϕ, ψ) such that

Theorem 5.3.

$$\mathcal{K}_M(u_1, v_1, \xi_{u_1}, \xi_{v_1}) = \mathcal{K}_M(I, J, \phi, \psi) = E' + B(\epsilon u_1, 0)\,[I + c(\epsilon J) + \mathcal{O}(\epsilon)] \ , \quad (5.9)$$

where the second derivative of the function $c(\epsilon J)$ does not vanish on an interval of the type $]A, +\infty[$.

The action variable I is, up to a multiplicative factor ϵ^{-2}, the variable I_u we defined previously. For the points of the motion situated on the magnetic field line, we recognize the magnetic moment $I = \frac{v_\perp^2}{2B}$ mentioned in the introduction. The second action variable is a function $J(c)$, which represents the area enclosed by the curve $\mathcal{C}_c$ defined as follows: $\frac{1}{2}\ \xi_{u_1}^2 - c\ B(\epsilon u_1, 0) \ = \ E'$. If we denote by B_M the minimal value of B on (L_M), $\mathcal{C}_c$ is a closed curve for $c \in]-\frac{E'}{B_M}, 0[$ (see Figures 5, 6).

We have

$$J(c) = \oint_{\mathcal{C}_c} \xi_{u_1} du_1 = \epsilon^{-1} \int_{u_{\min}}^{u_{\max}} \sqrt{2[E' + cB(u,0)]}du \,.$$

To obtain (5.9) it remains to check that $(\epsilon J)(c)$ is an increasing function on $]-\frac{E'}{B_M}, 0[$, and that its derivative is also increasing on an interval of the type $]a, 0[$. This is due to the asymptotic behaviour of $B(u)$ at infinity. On figure 6 we set $E' = B_M = 1$; it can be seen that the area enclosed by the curve $\mathcal{C}_c$ is increasing as c grows from $-0,9$ to $-0,1$.

The theorem (5.3) entails that $c''(\epsilon J)$ vanishes only on a finite number of values $J_1, \dots, J_p$. It is possible then to apply Moser's theorem on each annulus of the type $\mathcal{A}(\epsilon J_k, \epsilon J_{k+1})$ and on each annulus exterior to the circles $J = J_1$ and $J = J_p$.

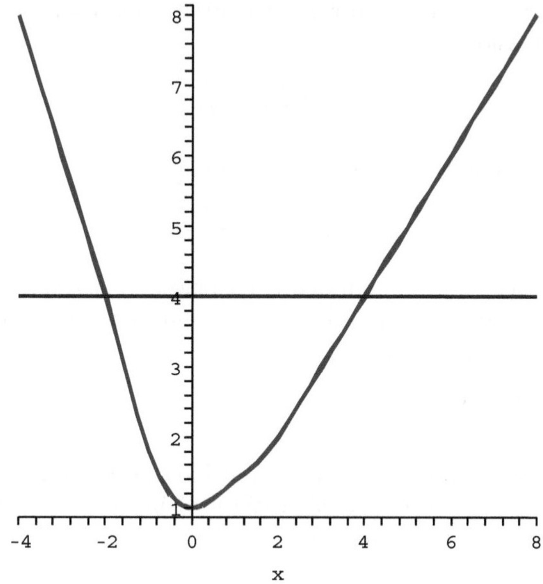

FIGURE 5. The function $B(u)$.

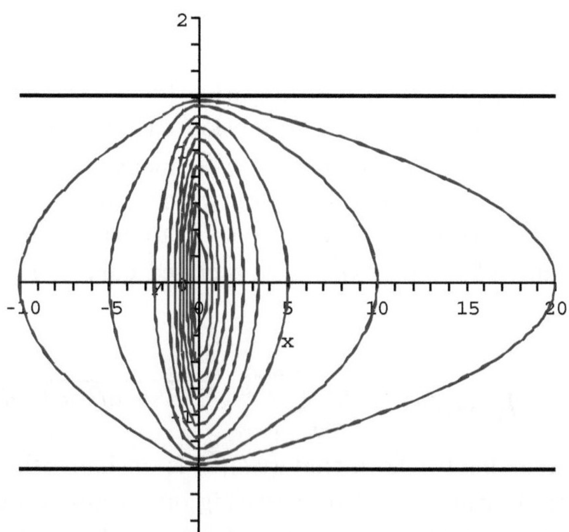

FIGURE 6. Curves $\mathcal{C}_c$.

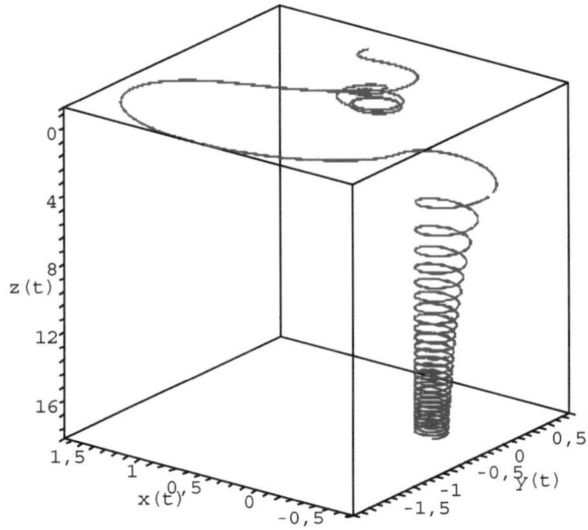

FIGURE 7. A trajectory.

On such an annulus the action J can be expressed as a function of the new time ϕ. One shows that the diffeomorphism of the annulus $(J(0), \psi(0)) \rightarrow (J(2\pi), \psi(2\pi))$ verifies Moser's condition (see [44], [45]). In fact this condition expresses the fact that $c''(\epsilon J)$ does not vanish. Let us notice that Moser's condition can be shown to be equivalent to the weak non-degeneracy condition introduced by Arnold in [2]. The interesting fact is that it is possible to check it explicitly in this setup.

Therefore we obtain the existence of an infinite number of curves which are invariant by this diffeomorphism. The curves generate invariant tori ; they foliate the surface of energy $\mathcal{H} = E$ so that any trajectory starting between two tori remains between those tori. Consequently the quantity I_u is a perpetual adiabatic invariant.

According to Remark 5 of the previous section we get then the following result:

Theorem 5.4. *Let M be fixed. There exist $\epsilon_0 > 0$ and $K > 0$ such that the trajectory of the solution is bounded, provided the following conditions are fulfilled*

$$(CI) \qquad E < \frac{\epsilon^2}{16}|M|^{2/3}.$$

$$(CI)' \qquad I_u(0) > 2K\epsilon^3.$$

for an $\epsilon < \epsilon_0$.

The conditions (CI) and $(CI)'$ are compatible. They express the fact that the velocities have to be small compared with positions whereas the component of

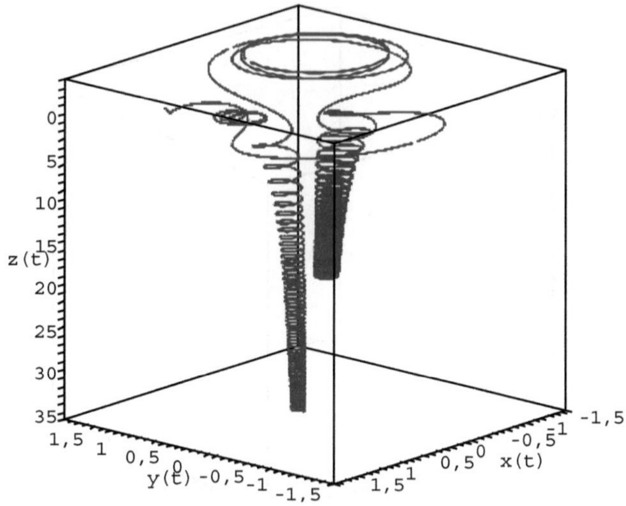

FIGURE 8. The same trajectory, continued.

the velocity normal to the magnetic field line has to be not too small compared with the total energy. In particular this excludes the case $z(0) = \dot{z}(0) = \dot{\theta}(0) = 0$, which corresponds to a particle starting on the magnetic field line (L_0) (defined by $\theta = \theta(0), z = 0$) with a velocity parallel to this line. Obviously in that case the motion is not bounded since the particle rolls up along the line (L_0), which can be seen as the bottom of the well for the degenerate potential $V(u, v) = B(u, 0)\ v^2$.

Figures 7 and 8 represent the motion of a particle computed by numerical simulation. The initial conditions are $M_0(1, 1, 0)$ and $V_0(0, 15; -0, 25; 0, 25)$. When the particle goes away from the origin, the strip $\mathcal{B}$ in which it is contained (considering a vertical section) clearly appears. The numerical simulation seems to suggest that the motion downwards is bounded: the radius of the helix decreases but the motion is stabilized and reflected and the particle is returning towards the origin to an other helicoidal-like motion.

6. Open problems and conclusion

- The last work presented gives of course only a partial answer. The symmetry argument is crucial, because the invariant tori generated by the unperturbed Hamiltonian prevent the particle to go away. In the non symmetric case the phase space has dimension 6, the energy surface has dimension 5 and the tori do not play a limitative role any longer. It would be of real interest to investigate if there exists a drift exponentially small for that situation, as suggested by the works of Nekhoroshev and Georgilli [47], [19]. There is a paper by G. Benettin and P. Sempio [4] which goes in that direction, requiring three time scales for such a motion.

- Concerning magnetic bottles in hyperbolic geometry, we gave a generalization to the geometrically finite hyperbolic surfaces, in the case when these manifolds are of infinite volume.

  Another generalization of the result would be to investigate the three-dimensional case. What properties should we require on the magnetic field to get a non-Weyl-type estimate in the hyperbolic half-space?

  Finally, from a more geometrical point of view it would be nice to understand what term is replacing the additional term $1/4$, which is a feature of the hyperbolic geometry, when another metric is considered.

- Another natural problem about magnetic bottles has been considered in [10].

  Let us consider a particle in a domain Ω in $\mathbb{R}^d$ ($d \geq 2$) in the presence of a magnetic field B. The topological boundary $\partial\Omega$ of Ω is assumed to be compact. At the classical level, if the strength of the field tends to infinity as x approaches the boundary $\partial\Omega$, the charged particle is expected to be confined and never visit the boundary: the Hamiltonian dynamics is complete. At the quantum level the fact that the particle never feels the boundary amounts to saying that the magnetic field completely determines the motion, so there is no need for boundary conditions. At the mathematical level, the problem is to find conditions on the behavior of $B(x)$ as x tends to $\partial\Omega$ which ensure that the magnetic operator H_A is essentially self-adjoint on $C_o^\infty(\Omega)$. These conditions will not depend on the gauge A, but only on the field B. This question may be of technological interest in the construction of tokamacs for the nuclear fusion [57]. The ionized plasma which is heated is confined thanks to magnetic fields.

  The result is the following: under some continuity assumption on the direction of $B(x)$ at the boundary, for any $\epsilon > 0$ and $R > 0$, there exists a constant $C_{\epsilon,R} \in \mathbb{R}$ such that, $\forall u \in C_o^\infty(\Omega)$, the quadratic form h_A satisfies the quite optimal bound

$$h_A(u) \geq (1-\epsilon) \int_{\Omega \cap \{x \mid |x| \leq R\}} |B|_{\text{sp}} \, |u|^2 \, |dx| - C_{\epsilon,R} \, \|u\|^2 \,. \tag{6.1}$$

Here $|B(x)|_{\text{sp}}$ is a suitable norm on the space of bi-linear antisymmetric forms on $\mathbb{R}^d$, called the *spectral norm*. This implies that H_A is essentially self-adjoint if $|B(x)|_{\text{sp}} \geq (1+\eta)D(x)^{-2}$ where $\eta > 0$ and D is the distance to the boundary of Ω.

Examples of such magnetic bottles are given in the following cases:

- The domain Ω is a polytope
- The boundary is smooth and the Euler characteristic vanishes (toroidal domain)
- The boundary is smooth and the Euler characteristic does not vanish (non toroidal domain)
- Monopoles and dipoles in $\Omega = \mathbb{R}^3 \setminus 0$

For any $\epsilon > 0$ and when Ω is the unit disk, an example of a non essentially self-adjoint operator H_A is given with $|B(x)|_{\rm sp} \sim (\sqrt{3}/2 - \epsilon)D(x)^{-2}$ showing that the previous bound is rather sharp.
The following questions seem to be quite interesting:
  - What are the properties of a *classical* charged particle in a confining magnetic box? Are almost all trajectories not hitting the boundary?
  - What is the *optimal* constant C in the estimates $|B(x)|_{\rm sp} \geq CD(x)^{-2}$? (We know that the optimal constant lies in the interval $[\sqrt{3}/2, 1]$.)
- In conclusion, we tried in this paper to highlight the relationship between magnetic bottles and degenerate potentials, as well in the classical mechanics context as in the quantum mechanics one. The Weyl asymptotics have to be revisited in both cases, and the classical Hamiltonian induced by a magnetic bottle can be seen as a perturbation of the Hamiltonian derived from an operator with a degenerate potential. It could be nice to go further in that comparison, by trying to express the non-Weyl-type asymptotics in a unified way.

References

[1] V.I. Arnold, *Dynamical systems.* Springer Verlag, Encyclopaedia of Math Sc **3** 1988.

[2] V.I. Arnold, *Small denominators and problems of stability of motion in classical and celestial dynamics.* Russ. Math. Survey **18, 6** (1963), 85–190.

[3] J. Avron, I. Herbst, B. Simon, *Schrödinger operators with magnetic fields.* Duke. Math. J **45** (1978), 847–883.

[4] G. Benettin, P. Sempio, *Adiabatic invariants and trapping of a point charge in a strong non-uniform magnetic field.* Nonlinearity **7** (1994), 281–303.

[5] M. Braun, *Particle motions in a magnetic field.* Journal of Diff. Equ. **8** (1970), 294–332.

[6] Y. Colin de Verdière, *L'asymptotique de Weyl pour les bouteilles magnétiques.* Comm. Math. Phys **105** (1986), 327–335.

[7] Y. Colin de Verdière, *Quasi-modes sur les variétés riemanniennes.* Invent. Math **43,1** (1977), 15–52.

[8] Y. Colin de Verdière, *L'asymptotique de Weyl pour les bouteilles magnétiques bi-dimensionnelles.* Prépublications de l'Institut Fourier **33** (1985).

[9] Y. Colin de Verdière *Minorations de sommes de valeurs propres d'un domaine et conjecture de Polya.* Séminaire de Théorie spectrale et Géométrie (1984–85).

[10] Y. Colin de Verdière, F. Truc, *Confining quantum particles with a purely magnetic field.* (2009), http://hal.archives-ouvertes.fr/hal-00365828/en/, to appear in Annales de l'Institut Fourier.

[11] A. Comtet, *On the Landau Levels on the hyperbolic space.* Ann. Phys **173** (1987), 185–209.

[12] J.S. de Wet and Mandl, *On the asymptotic distribution of eigenvalues.* Proc. Roy. Soc. London Ser. **200** (1950), 572–580.

[13] S. Doi, A. Iwatsuka, T. Mine, *The uniqueness of the integrated density of states for the Shrodinger operators with magnetic fields.* Mathematische Zeitschrift **237** (2001), 335–371.

[14] A. Dufresnoy, *Un exemple de champ magnétique dans* $\mathbb{R}^\nu$. Duke. Math. J. **50** (1983), 729–734.

[15] J. Elstrodt, *Die Resolvente zum Eigenwertproblem der automorphen Formen in der hyperbolischen Ebene I, II, III.* Math. Ann., **203** (1973) 295–330, Math. Z. **132** (1973) 99–134, Math. Ann. **208** (1974) 99–132.

[16] S. Fournais, B. Helffer, *Accurate eigenvalue asymptotics for the magnetic Neumann Laplacian.* Annales de l'Institut Fourier **56,1** (2006), 1–67.

[17] S. Fournais, B. Helffer, *On the third critical field in Ginzburg-Landau theory.* Comm. Math. Phys. **266 (1)** (2006), 153–196.

[18] M. Gardner, T*he adiabatic invariant of periodic classical systems.* Phys.Rev. **115** (1959), 791–794.

[19] A. Giorgilli, *Rigorous results on the power expansions for integrals of a hamiltonian system near an elliptic equilibrium.* Ann. Inst. Henri Poincaré **48 (4)** (1988), 423–439.

[20] S. Golénia, S. Moroianu, *Spectral analysis of magnetic Laplacians on conformally cusp manifolds.* arχiv:math0707780v4, 2007.

[21] C. Grosche, *The path integral on the Poincaré upper half-plane with magnetic field and for the Morse potential.* Ann. Phys. **187** (1988), 110–134.

[22] B. Helffer, A. Morame, *Magnetic bottles in connection with superconductivity.* J. of Functional Anal. **185** (2001), 604–680.

[23] B. Helffer, A. Morame, *Magnetic bottles for the Neumann problem: the case of dimension* 3. Proc. Indian Acad. Sci. **112 (1)** (2002), 71–84.

[24] B. Helffer, A. Morame, *Magnetic bottles for the Neumann problem: curvature effects in the case of dimension* 3 (*general case*). Ann. Sc. Ec. Norm. Sup. **37 (4)** (2004), 105–170.

[25] L. Hörmander, *Hypoelliptic second order differential equations* Acta. Math. **119** (1967), 147–171.

[26] N. Ikeda, *Brownian Motion on the Hyperbolic plane and Selberg Trace Formula.* J. Func. Anal. **163** (1999), 63–110.

[27] Y. Inahama, S. Shirai, *The essential spectrum of Schrödinger operators with asymptotically constant magnetic fields on the Poincaré upper half-plane.* J. Math. Phys. **44** (2003), 89–106.

[28] Y. Inahama, S. Shirai, *Eigenvalue asymptotics for the Schrödinger operators on the hyperbolic plane.* J. Func. Anal. **211** (2004), 424–456.

[29] Y. Inahama, S. Shirai, *Spectral properties of Pauli operators on the Poincaré upper half-plane.* J. math. Phys. **44** (2003), 2451–2462.

[30] A. Iwatsuka, *Magnetic Schrödinger operators with compact resolvent.* J. Math. Kyoto. Univ. **26 (3)** (1986), 357–374.

[31] A.G. Kostjucenko, *Asymptotic distribution of the eigenvalues of elliptic operators.* Soviet Math. Dokl. **5** (1964), 1171–1175.

[32] M. Kruskal, *Asymptotic theory of Hamiltonian and other systems with all solutions nearly periodic.* Journal of Math. Phys. **3** (1962), 806–829.

[33] B.M. Levitan, *On the asymptotic behavior of Green's function and its expansion in eigenvalues of Schrödinger's equation.* Math. USSR-Sb. **41, 83** (1957), 439–458.

[34] K. Lu, X-B. Pan, *Estimates of the upper critical field for the Ginzburg-Landau equations of superconductivity.* Physica **D 127** (1999), 73–104.

[35] K. Lu, X-B. Pan, *Eigenvalue problems of Ginzburg-Landau operator in bounded domains.* Journal of Math. Physics **40 (6)** (1999), 2647–2670.

[36] K. Lu, X-B. Pan, *Surface nucleation of superconductivity in* 3*-dimension.* J. of Differential Equations 1**68 (2)** (2000), 386–452.

[37] A. Martinez, *Développement asymptotiques et effet tunnel dans l'approximation de Born-Oppenheimer.* Ann. Inst. Henri Poincaré **49, (3)** (1989), 239–257.

[38] H. Matsumoto, *Semiclassical asymptotics of eigenvalue distributions for Schrödinger operators with magnetic fields.* Comm. in Partial Diff. Eq. **19** (1994), 719–759.

[39] A. Morame and F. Truc, *Semiclassical Eigenvalue Asymptotics for a Schrödinger Operator with Degenerate Potential.* Asymptotic Anal. **22, (1)** (2000), 39–49.

[40] A. Morame, F. Truc, *Remarks on the spectrum of the Neuman problem with magnetic field in the half-space.* Journal of Mathematical Physics **46, (1)** (2005), 1–13.

[41] A. Morame, F. Truc, *Accuracy on eigenvalues for a Schrödinger Operator with a Degenerate Potential in the semi-classical limit.* Cubo, A Mathematical Journal **9, (2)** (2007), 1–14.

[42] A. Morame, F. Truc, *Magnetic bottles on the Poincaré half-plane: spectral asymptotics;* Journal of Mathematics of Kyoto University **48, 3** (2008)

[43] A. Morame, F. Truc, *Magnetic bottles on geometrically finite hyperbolic surfaces* Journal of Geometry and Physics, **59** (2009), 1079–1085.

[44] J. Moser, *On invariant curves of area preserving mappings of an annulus.* Nachr. Acad. Wiss. II Göttingen, Math. Phys. Klasse, (1962), 1–20.

[45] J. Moser, *Stable and random motions in dynamical systems.* Annals of math. studies. Princeton University Press, (1973), 1–20.

[46] A.I. Neistadt, *The separation of motions in systems with rapidly rotating phase;* J. Appl. Math. Mech. **48** (1984), 133–139.

[47] N.N. Nekhoroshev, *An exponential estimate of the time of stability of nearly-integrable hamiltonian systems.* Russ. Math. Surveys **32** (1977), 1–65.

[48] T.G. Northrop, *The adiabatic motion of charged particles.* Wiley Interscience Publishers New York, 1963.

[49] M. Reeds, B. Simon, *Methods of Modern Mathematical Physics.* Academic Press, New York, 1978.

[50] D. Robert, *Comportement asymptotique des valeurs propres d'opérateurs du type de Schrödinger à potentiel dégénéré.* J. Math. Pures et Appl. **61** (1982), 275–300.

[51] G.V. Rozenbljum, *Asymptotics of the eigenvalues of the Schrödinger operator.* Math. USSR Sbornik **22, (3)** (1974), 349–371.

[52] B. Simon, *Nonclassical eigenvalue asymptotics.* J. of Funct. Analysis **53** (1983), 84–98.

[53] M.Z. Solomyak, *Asymptotics of the spectrum of the Schrödinger operator with non-regular homogeneous potential.* Math. USSR Sbornik **55, (1)** (1986), 19–37.

[54] M. Shubin, *The essential Self-adjointness for Semi-bounded Magnetic Schrödinger operators on Non-compact Manifolds.* J. Func. Anal. **186** (2001), 92–116.

[55] H. Tamura, *Asymptotic distribution of eigenvalues for Schrödinger operators with magnetic fields.* Nagoya Math. J. **105** (1987), 40–69.

[56] E.C. Titchmarsh, *On the asymptotic distribution of eigenvalues.* Quart. J. Math. Oxford Ser. **2, 5** (1954), 228–240.

[57] http://en.wikipedia.org/wiki/Tokamak

[58] F. Truc, *Trajectoires bornées d'une particule soumise à un champ magnétique linéaire.* Annales de l'IHP (Physique théorique) **64** (1996), 127–154.

[59] F. Truc, *Semi-classical asymptotics for magnetic bottles.* Asympt. Anal. **15** (1997), 385–396

[60] F. Truc, *Born-Oppenheimer-type approximations for a degenerate potential: recent results and a survey on the area.* Operator theory: Advances and applications. **186** Birkhäuser Verlag, 403–413.

Françoise Truc
Institut Fourier
UMR 5582 CNRS-UJF
B.P. 74
F-38402 St Martin d'Hères Cedex, France
e-mail: `Francoise.Truc@ujf-grenoble.fr`

List of Participants

Al-Rawashdeh, Ahmed
Behncke, Horst
Behrndt, Jussi
Boutet de Monvel, Anne
Brown, Malcolm
Chelkak, Dmitry
Chernyshev, Vsevolod
Cichon, Dariusz
Clark, Stephen
Cojuhari, Alexandra
Combescure, Monique
Costa, Emanuele
Damanik, David
Del Rio, Rafael
Derevyagin, Maksym
Derezinski, Jan
Dombrowski, Joanne
Enblom, Andreas
Evans, W. Desmond
Exner, Pavel
Fournais, Søren
Frank, Rupert
Georgescu, Vladimir
Gesztesy, Fritz
Gilbert, Daphne
Janas, Jan
Kadem, Abdelouahab
Karpeshina, Yulia
Karwowski, Witold
Kheifets, Alexander
Kiselev, Alexander
Korotyaev, Evgeny
Kounkou, Roger
Krueger, Helge
Kurasov, Pavel
Langer, Matthias
Laptev, Ari
Luger, Annemarie
Malejki, Maria
Mikhaylov, Victor
Moszynski, Marcin
Motyka, Wojciech
Mykytyuk, Yaroslav
Naboko, Sergey
Nachtergaele, Bruno
Nowaczyk, Marlena
Parnovski, Leonid
Pashkova, Iuliia
Pavlov, Boris
Peller, Vladimir
Post, Olaf
Romanov, Roman
Rozenblioum, Grigori
Seiringer, Robert
Shkalikov, Andrei
Silva, Luis
Simonov, Sergey
Smilansky, Uzy
Stollman, Peter
Stolz, Gunter
Szwarc, Ryszard
Teschl, Gerald
Tikhonov, Alexey
Toft, Joachim
Tomilov, Yuri
Truc, Françoise
Trunk, Carsten
Veselic, Ivan
Watanabe, Kazuo
Wood, Ian G.
Yafaev, Dmitri
Yuditskii, Peter
Zielinski, Lech
Zygmunt, Marcin

Birkhäuser www.birkhauser-science.com

Operator Theory: Advances and Applications (OT)

This series is devoted to the publication of current research in operator theory, with particular emphasis on applications to classical analysis and the theory of integral equations, as well as to numerical analysis, mathematical physics and mathematical methods in electrical engineering.

Edited by
Joseph A. Ball (Blacksburg, VA, USA), Harry Dym (Rehovot, Israel), Marinus A. Kaashoek (Amsterdam, The Netherlands), Heinz Langer (Vienna, Austria), Christiane Tretter (Bern, Switzerland)

■ **OT 216: Ruzhansky, M. / Wirth, J.**, Modern Aspects of the Theory of Partial Differential Equations (due 2011).
ISBN 978-3-0348-0068-6

■ **OT 215: Cruz-Uribe, D. / Martell, J.M. / Perez, C.**, Weights, Extrapolation and the Theory of Rubio de Francia (due 2011).
ISBN 978-3-0348-0071-6

■ **OT 214: Rodino, L. / Wong, M.W. / Zhu, H. (eds.)**, Pseudo-Differential Operators: Analysis, Applications and Computations (due 2011).
ISBN 978-3-0348-0048-8

■ **OT 213: Janas, J. / Kurasov, P. / Laptev, A. / Naboko, S. / Stolz, G. (eds.)**, Spectral Theory and Analysis. Operator Theory, Analysis and Mathematical Physics (OTAMP) 2008, Bedlewo, Poland (2011).
ISBN 978-3-7643-9993-1

■ **OT 212: Curto, R.E. / Mathieu, M.**, Elementary Operators and Their Applications (2011).
ISBN 978-3-0348-0036-5

■ **OT 211: Demuth, M. / Schulze, B.-W. / Witt, I. (eds.)**, Partial Differential Equations and Spectral Theory (2010).
Subseries Advances in Partial Differential Equations.
ISBN 978-3-0346-0508-3

■ **OT 210: Duduchava, R. / Gohberg, I. / Grudsky, S.M. / Rabinovich, V. (eds.)**, Recent Trends in Toeplitz and Pseudodifferential Operators. The Nikolai Vasilevskii Anniversary Volume (2010).
ISBN 978-3-0346-0547-2

■ **OT 209: Eisner, T.**, Stability of Operators and Operator Semigroups (2010).
ISBN 978-3-0346-0194-8

■ **OT 208: Elin, M. / Shoikhet, D.**, Linearization Models for Complex Dynamical Systems (2010).
Subseries Linear Operators and Linear Systems.
ISBN 978-3-0346-0508-3

■ **OT 207: Axler, S. / Rosenthal, P. / Sarason, D. (eds.)**, A Glimpse at Hilbert Space Operators. Paul R. Halmos in Memoriam (2010).
ISBN 978-3-0346-0346-1

■ **OT 206: Lerer, L. / Olshevsky, V. / Spitkovsky, I.M. (eds.)**, Convolution Equations and Singular Integral Operators (2010).
ISBN 978-3-7643-8955-0

■ **OT 205: Schulze, B.-W. / Wong, M.W. (eds.)**, Pseudo-Differential Operators: Complex Analysis and Partial Differential Equations (2009).
ISBN 978-3-0346-0197-9

■ **OT 204: Frazho, A. / Bhosri, W.** An Operator Perspective on Signals and Systems (2009).
Subseries Linear Operators and Linear Systems.
ISBN 978-3-0346-0291-4

■ **OT 202/203: Ball, J.A. / Bolotnikov, V. / Helton, J.W. / Rodman, L. / Spitkovsky, I.M. (eds.)**, Topics in Operator Theory. A Tribute to Israel Gohberg on the Occasion of his 80th Birthday (2010).
Volume 1: Operators, Matrices and Analytic Functions. ISBN 978-3-0346-0157-3
Volume 2: Systems and Mathematical Physics.
ISBN 978-3-0346-0160-3
Set Volumes 1/2: ISBN 978-3-0346-0163-4

MIX
Papier aus verantwortungsvollen Quellen
Paper from responsible sources
FSC® C105338

If you have any concerns about our products,
you can contact us on
ProductSafety@springernature.com

In case Publisher is established outside the EU,
the EU authorized representative is:
Springer Nature Customer Service Center GmbH
Europaplatz 3, 69115 Heidelberg, Germany

Printed by Libri Plureos GmbH
in Hamburg, Germany